AF579986

Otherwhere Ethnography

Otherwhere Ethnography

An Introduction to Outer Space Studies

Edited by

ISTVAN PRAET
PERIG PITROU

OXFORD
UNIVERSITY PRESS

OXFORD
UNIVERSITY PRESS

Oxford University Press is a department of the University of Oxford.
It furthers the University's objective of excellence in research, scholarship,
and education by publishing worldwide. Oxford is a registered trade mark of
Oxford University Press in the UK and in certain other countries.

Published in the United States of America by Oxford University Press
198 Madison Avenue, New York, NY 10016, United States of America.

Library of Congress Cataloging-in-Publication Data

Names: Anthropology off Earth Conference (2019 : Paris, France) author |
Praet, Istvan, 1974- editor | Pitrou, Perig, 1973- editor
Title: Otherwhere ethnography : an introduction to outer space studies /
edited by Istvan Praet and Perig Pitrou.
Description: New York, NY, United States of America : Oxford University
Press, [2025] | Includes bibliographical references and index.
Identifiers: LCCN 2025015900 | ISBN 9780197790854 hardback |
ISBN 9780197790878 epub | ISBN 9780197790885
Subjects: LCSH: Outer space—Exploration—Philosophy—Congresses |
Manned space flight—Philosophy—Congresses | Human geography—Congresses |
Astrogeology—Congresses | LCGFT: Conference papers and proceedings
Classification: LCC QB495 .A56 2019 | DDC 919.904—dc23/eng/20250411
LC record available at https://lccn.loc.gov/2025015900

DOI: 10.1093/9780197790885.001.0001

Printed by Marquis Book Printing, Canada

The manufacturer's authorized representative in the EU for product safety is
Oxford University Press España S.A., Parque Empresarial San Fernando de Henares,
Avenida de Castilla, 2 – 28830 Madrid (www.oup.es/en).

Contents

PART II OTHERWHERE ETHNOGRAPHIES: ANALOGS, INSTRUMENTS, ARTIFACTS, VIRUSES, AND VEGETABLES

Acknowledgements

This collection is the result of a sustained collaboration that originated at the 2019 Anthropology Off Earth conference, which the editors co-organized together with Régis Ferrière, Joffrey Becker, and Elsa De Smet at the Collège de France and the Observatoire de Paris. The conference was funded by the IRIS/OCAV/Paris Sciences et Lettres research consortium, and we thank Stéphane Mazevet and Ludovic Jullien for supporting us without fail. An international workshop at the CNRS-ENS-UA International Research Center iGLOBES, which the editors co-organized with Ferrière and Kevin Bonine at the University of Arizona later that year, enabled us to refine our ideas: We thank John Adams, chief operations officer of the Biosphere 2 facility, and Joaquin Ruiz, its director, for their support. Our volume acquired its final shape at the 2022 Ethnographies of Outer Space conference, co-organized at the University of Trento by Praet and Valentina Marcheselli, whom we thank warmly. Thanks are also due to Denis Sivkov, Lauren Reid, Paola Castaño, Juan Francisco Salazar, and Joffrey Becker. Their ideas and suggestions have influenced this volume in a number of ways, even though they did not contribute a chapter of their own in the end. We furthermore thank Federico De Musso for designing the cover image and for kindly providing us several additional images used in this book. Finally, we are very grateful to the marvelously effective Maartje Scheltens, who helped us greatly to rework and refine our initial book proposal.

List of Contributors

Victor Buchli, PhD Professor of Material Culture
Department of Anthropology
University College London
London, UK

Luis A. Campos, PhD Baker College Chair for the History of Science, Technology, and Innovation
History Department
Rice University
Houston, TX, USA

Davide Chinigò Senior Assistant Professor
University of Bologna
Bologna, Italy

Stefan Helmreich Elting E. Morison Professor of Anthropology
Department of Anthropology
Massachusetts Institute of Technology
Cambridge, MA, USA

David Jeevendrampillai Lecturer in Anthropology
The University of Manchester
Manchester, UK

Valentina Marcheselli Postdoctoral Research Fellow
Department of Philosophy and Cultural Heritage
Ca' Foscari University of Venice
Venice, Italy

Lisa Messeri, PhD Associate Professor
Department of Anthropology
Yale University
New Haven, CT, USA

Zara Mirmalek, PhD VIPER Mission Deputy Science Operations and Integration
NASA Ames Research Center
Moffett Field, CA, USA
Senior Social Scientist
BAERI
Moffett Field, CA, USA

Valerie Olson, PhD Associate Professor
University of California, Irvine
Irvine, CA, USA

Aaron Parkhurst, PhD Associate Professor in Biosocial and Medical Anthropology
Department of Anthropology
University College London
London, UK

Perig Pitrou CNRS Research Director
Laboratoire d'anthropologie sociale
Paris, France and
Maison Francaise d'Oxford
Oxford University
Oxford, United Kingdom

Istvan Praet Lecturer in Anthropology
Department of Anthropology
Durham University
Durham, United Kingdom

David Valentine Associate Professor
Department of Anthropology
University of Minnesota Twin Cities
Minneapolis, MN, USA

Cherryl Walker, DLitt Emeritus Professor
Department of Sociology and Social Anthropology
Stellenbosch University
Stellenbosch, South Africa

Introduction

Otherwhere Ethnography, a Means to Reimagine Contemporary Space Exploration

Perig Pitrou and Istvan Praet

A Cargo Cult of Worldwide Proportions

More than half a century into the Space Age, contemporary space exploration remains steeped in a distinctive kind of naive optimism. By and large, it continues to be closely associated with the advance of high-tech and rocket engineering, and with an unshakable belief in unidirectional scientific progress. Space exploration tends to be framed as a tremendous spectacle one is supposed to stare at in awe, open-mouthed. There is excitement and suspense—parachuting a probe on Titan, flying a robotic helicopter on Mars—endless novelties—the newly deployed James Webb telescope's mesmerizing portrait of a ghostly, whitish Neptune and its rings—and moments of high drama—the death of Laika the space dog, the *Challenger* and *Columbia* Space Shuttle disasters. The aim of this book is neither to crush this basic sense of wonder and risky adventure nor to downplay the brilliant feats of engineering that undergird off-Earth exploration. But our contributors do formulate important qualifications and serious reservations. What we present here is a collection of "otherwhere ethnographies," richly detailed accounts of how space research and space enterprises are being rethought in an age in which extraterrestrial exploration is no longer the monopoly of a handful of superpowers (with thanks to David Valentine, who coined the term).

Each in their own way, the otherwhere ethnographers who contribute to this volume question the scientistic attitudes that prevail in the space professions, whether public or private, academic or commercial. They show that scientism, the conviction that "modern rationality" and "the modern scientific method" are the only way to render truth about the cosmos, is not only deeply problematic but ultimately also ill-suited to our times. For one thing,

Perig Pitrou and Istvan Praet, *Introduction*. In: *Otherwhere Ethnography*. Edited by: Istvan Praet and Perig Pitrou, Oxford University Press. © Oxford University Press (2025). DOI: 10.1093/9780197790885.003.0001 the scientistic attitude has failed to come to terms with the manifold connections between modern forms of thought, productivism, and colonialism. Scientism, embedded in discourses of frontiers and discovery, remains blind to the fact that the contemporary exploration of outer space has a rather less glorious side to it as well. As many of our contributors underline, there are good arguments to view it as just another instance of neocolonialism, as an enterprise with the potential of expanding rather than mitigating the ravages of predatory capitalism and terrestrial geopolitics (even though Valentine, in particular, warns us in Chapter 4 that this view is also reductive). It is telling that the current flag bearers of scientism and the concomitant belief in cumulative progress are not so much high-profile academics or famous intellectuals as a trio of Anglo-American space billionaires. Elon Musk, Jeff Bezos, and Richard Branson have become veritable gurus for a huge following of space entrepreneurs and space researchers, and they are hailed as heroic pioneers by a global public of space enthusiasts and media acolytes. Many seem to look up to them as latter-day prophets.

Anthropologically speaking, this is of course fascinating. One is tempted to picture them—to co-opt an image from the anthropologist Gísli Pálsson (2007: 45)—as the "big men" at the center of a Melanesian cargo cult of worldwide proportions; just as the Papuan chiefs of old who promised the imminent arrival of planes full of luxuries and valuable commodities, Musk, Bezos, and Branson announce the impending occurrence of amazing things. *Homo sapiens*, it is foretold, will soon be a biplanetary species. Asteroid mining and the exploitation of lunar resources will lead to a global economic boom and are bound to create unimaginable riches and opportunities. The slogans and the corporate pitches of their respective space companies speak for themselves: "Making life multiplanetary" and "Advancing humanity's ability to live and work in space" (SpaceX); "a vision of millions of people living and working in space for the benefit of Earth" and "a time when people can tap into the limitless resources of space and enable the movement of damaging industries into space to preserve Earth, humanity's blue origin" (Blue Origin); and "to connect people across the globe to the love, wonder and awe created by space travel" (Virgin Galactic). Varda Space Industries, an in-orbit manufacturing company funded by the venture capitalist and Silicon Valley figure Peter Thiel, summarizes its mission as "expanding the economic bounds of humankind" and "[to] be at the helm of an orbital industrial revolution." With a similar sense of understatement, aerospace company Lockheed Martin states on its corporate website,

> We go for the next generation. We go for progress. We go for a greater understanding of the universe. We go to prove that getting to Mars is possible by using *today's* technology. . . . At Lockheed Martin, we go for the greatest adventure in human history.

On X (Twitter), OpenAI CEO Sam Altman has put it more succinctly: Just as we can cure all human disease, "we can colonize space."

The central aim of this book is not so much to critique these scientistic, military–industrial dreams and neocolonial visions formulated with such earnest pomposity (although Stefan Helmreich has a good stab at it in Chapter 10) as to reveal their provincial and curiously uniform nature. Rather than exposing the hubris, the greed, or any other possible motives behind such grandiose schemes, our contributors proceed in a more indirect way. Essentially, they demonstrate that space exploration can also be thought otherwise. The Germans have a beautiful if untranslatable expression for the kind of thinking epitomized by the aforementioned space companies: *Einheitswurst*—literally, a uniform sausage. If this volume has one merit, it must be that it looks beyond this singular epistemic sausage and its neoliberal apple sauce and that it introduces the possibility of other epistemic flavors and indeed other outer spaces. This is what Lisa Messeri and Valerie Olson refer to in Chapter 1 as the pluralization of space exploration: the recognition that "outer space" must be grasped as a plural space, for social groups as well as analysts and professional researchers. In other words, outer space cannot and should not be reduced to a privileged playground for an elite bunch of egomaniacal space barons and their bleary-eyed devotees. One must resist viewing it as just another frontier for extractive capitalism, or as a mere chessboard on which the imperial ambitions and/or escape phantasies of fanatical tech elites and Silicon Valley preppers take shape.

Space Age Modernity as a World on the Wane

The philosopher and sociologist of science Bruno Latour relished to emphasize that the Mars settlement plans of Musk and company are *already* hopelessly outdated (Latour and Weibel 2020, passim). Some may find it exaggerated, and needlessly provocative, to describe the leading rocket entrepreneurs of our day as painfully old-fashioned, and a little boring to boot. But Latour does this in a very precise sense. What he is getting at, basically, is that Musk, Bezos, Branson, and their cheerleaders conceive of space exploration in an archetypically *modern* way. Now for Latour modernity is a world on the wane rather than a beckoning horizon or a promising future. In a sense, the modern epoch is already a thing from the past, rather like historical periods such as the Renaissance or the Baroque—that is why he speaks of "the modern parenthesis." The modern period has closed, it is just that some are still in denial or have simply failed to notice it (cf. Danowski and Viveiros de Castro 2016, Descola and Pignocchi 2022). From this vantage point, contemporary space billionaires are the last diehard representatives of a metaphysical tradition inaugurated in the 17th century by the likes of Galileo, Descartes, and Bacon. And while Latour is the first to recognize that this tradition has great merits that historians of science have rightly celebrated, it is of limited use if one seeks to come to terms with the predicament of the inhabitants of a planet facing global warming and a major ecological crisis. For all their glib talk about inspiring future generations and tackling terrestrial challenges by reaching for the stars, the conceptual luggage carried by companies such as SpaceX and Blue Origin is century-old, hyper-conservative, and badly adapted to the circumstances of the Anthropocene.

Although the contributors of this volume may disagree about some of the finer points of this general analysis—some fear that Latour's announcement of the closure of modernity might be an instance of wishful thinking rather than a true reflection of the daunting future that awaits us—it is fair to say that they all agree with its broad brushstrokes. All subscribe to the generic idea that modern thinking in its current Space Age crystallization ultimately boils down to a type of intellectual colonialism. All are possessed by the eerie sense that the seemingly unstoppable drive to exploit resources on and beyond Earth and the scientistic push to extend the "edge of objectivity" ever further and further are two sides of the same coin. None disagree with Campos' observation in Chapter 2 that "the history of terrestrial politics is an inescapable feature of our understandings of extraterrestrial elsewheres." All share the broad ambition to decolonize space exploration and to rethink it in ways that are not so much post-modern—for many extraterrestrial endeavors manifestly remain configured in quintessentially modern ways—as extra-modern. Hence the blossoming of multiple alternative perspectives in this volume and in the emerging field of outer space studies more generally: feminist (Holbrook et al. 2018, Klinger 2021), queer (Campos n.d.), non-Western (Chinigò and Walker, Chapter 5, this volume; Reid 2023), amateur (Sivkov n.d.), artistic (Becker 2018), religious (Traphagan 2020), and indigenous (Chang 2016, Casumbal-Salazar 2017). Challenging presentist habits of thought, David Valentine (Chapter 4, this volume) even introduces us to the vantage point of future human settlers on Mars. Nonhuman, other-than-human, and more-than-human perspectives are burgeoning too: integrated human–robot teams (Vertesi 2015; Mirmalek 2020 and Chapter 6, this volume); archaeological and object-oriented (Gorman 2019; Buchli, Chapter 9, this volume); image-centered (De Smet 2018); the interweaving of technical and vital processes, of artifacts and organisms (Pitrou 2019, Becker n.d.); botanical (Parkhurst and Jeevendrampillai, Chapter 7, this volume); microbial/viral (Helmreich 2009, 2016, and Chapter 10, this volume); and the perspectives mediated by various systems, models, and analogues (Olson 2018; Pitrou 2020, Marcheselli, Chapter 8, this volume; Mirmalek, Chapter 6, this volume).

This proliferation of multiple angles further implies that the so-called Space Age, conventionally understood as a post–World War II affair, can only be fully grasped if one situates it in a much more comprehensive history of exploration that remains to be written, as Luis Campos indicates in his pathbreaking account of Soviet astrobotany in Chapter 2 and Istvan Praet also alludes to in Chapter 3, which features both Renaissance seafarers and latter-day spacefarers. This broader take necessarily includes non-Western and Indigenous histories of exploration as well: What remains undertheorized is that Polynesian canoes or Hopi ladders are not in any absolute sense inferior to the latest space telescopes or state-of-the-art rockets. Star-based navigation between Pacific islands or ladder-based interaction with *kachina* star-beings no doubt have a great deal of untapped conceptual potential for latter-day space researchers. The harpoons of the Chachi spear fishers of Esmeraldas (Ecuador) are not so much weapons to hunt fishes as instruments to engage with underwater aliens (Praet 2013). The river surface, one might say, is their Kármán line, while the occasional drowning is conceived of as a form of alien abduction. And the harpoons themselves are perhaps best compared to the lightsabers we know from science fiction. When one reads Amazonian anthropology in this way, it becomes evident that the talent to build ingenious contraptions to explore alien worlds has never been the exclusive birthright of Westerners. As Perig Pitrou likes to emphasize, NASA engineers still have much to learn from the humble discipline known as the anthropology of techniques. In an age in which increasingly more astrobiology projects are conducted by means of terrestrial analogue sites, the Space Age obsession with jet propulsion and gleaming rockets may—who knows—just be a phase.

The Interplanetary Medium, a Modern Fetish

At any rate, there is an emerging consensus among scholars in outer space studies that there is something deeply strange about the current global fixation with extraterrestrial exploration, and about the modern assumption of a fundamental division between a planetary home and a not-so-homey elsewhere. On the one hand, moderns recognize one familiar planet—Earth. On the other hand, they marvel about a great number of celestial bodies they qualify as alien: planets such as Venus or Mars, moons such as Io or Europa, but also comets, asteroids, and much else. The interplanetary medium functions as their primary in/out partition—as their boundary of boundaries. Yet a discipline such as anthropology has very little to say about it: Although its role of separating "here" from "elsewhere" is pivotal in modern cosmology, the interplanetary medium remains poorly theorized. Even in the humanities and in social scientific research more widely, there exists no contemporary equivalent of Edward Grant's (1981) unsurpassed *Much Ado About Nothing*, which focuses on theories of space and vacuum in the premodern period. Why has a boundary of such central significance been ignored so systematically in the social sciences?

The modern belief that the interplanetary medium has no equivalent anywhere, that it constitutes the one and only final frontier, seems ineradicable. All other boundaries are conceived of as secondary by default. Consider the Earth's principal "layers": crust, mantle, and core. The terrestrial core may be more difficult to reach than Pluto, but that does not make the mantle an equally formidable boundary as the medium of outer space. Because the core is defined as "inner" on a priori grounds, the mantle can only be conceived of as an in/in subdivision. This highlights one of the odder features of modern cosmology: The celestial sphere is attributed an "outer" quality that the lower lithosphere does not possess. "Up there" is the boundary that really matters, and that has really captured the modern imagination. "Down here" things are distinctly less interesting—for one thing, one will not encounter aliens there (but see Aït-Touati et al. [2022)] for an account that inverts this traditional perspective).

In fact, all terrestrial boundaries are secondary boundaries in the modern mindset. Fault lines between tectonic plates may be impressive, but they are not on a par with the interplanetary medium. When one walks across the San Andreas fault, one does not arrive at an alien plate. One merely passes from one plate onto another deeply similar, unflappably terrestrial plate. Wallace's line, which famously separates the Asian from the Australian faunas, is another example: It marks a notable discontinuity, to be sure, but that discontinuity is always conceived of as partial, never as total. Tigers and rhinos are very much unlike kangaroos and wombats, but all of them are envisaged as fellow living beings at the end of the day. Wombats, modern people believe, are not aliens. Oceans that separate continents are also conceived of as lesser boundaries: When you cross an ocean, you may reach a different continent but not an alien world. Under the rules of modern metaphysics, seafaring never has quite the same cachet as spacefaring: Only astronauts are entitled to encounter aliens, whereas seafarers—or any earthbound explorers for that matter—are constricted to discovering more of the same (other humans, other animals, other plants).

Provided that one thinks like a modern, all of this appears perfectly self-evident. But one of the central goals of scholars in the emerging field of outer space studies is to question this self-evidence and to challenge the uniquely privileged position that the interplanetary medium has—for no obvious reason whatsoever when one comes to think of it—been attributed. What distinguishes them from other scholars is their awareness that there are, in fact, many signs that the conventional metaphysical framework premised on one major boundary and multiple minor boundaries is beginning to crumble across the natural *and* the social sciences. In fields as disparate as astrobiology, the anthropology of science, and parrot research there is a growing sense that geological and biological boundaries do not necessarily have to be thought as minor boundaries. As a boundary, the interplanetary medium may well be overrated. However enthralling it may be, outer space is perhaps not the unique frontier it is so commonly assumed to be. After all, isn't it peculiar and downright inconsistent that Space Age moderns have no temporal equivalent for it? They believe in outer space but don't believe in outer time. It is worth to dwell on these general observations for a moment.

The Proliferation of Outer Spaces in the Early 21st Century

The astrobiologist Nathalie Cabrol, who directs the Carl Sagan Center at the SETI Institute, routinely speaks about past geological epochs as if they were different planets. She is a world expert on ancient martian lakes and has led expeditions to study various Mars analogue sites across South America.

As one travels from the low-altitude Atacama Desert to the high Andes, she explains in a recent interview, one experiences what she describes as the substitution of space for time (Cabrol 2022). As one gains altitude, humidity increases but the atmosphere decreases, and ultraviolet radiation increases. By playing with these physical parameters, one can imagine oneself on Mars in subsequent geological epochs. Cabrol likes to emphasize that Mars, just like Earth, must not be grasped as one continuous planet but, rather, as a succession of different planets: In the past, circumstances were so radically different that one cannot possibly recognize it as the same planet. For all intents and purposes, crossing a geological boundary *is* crossing a planetary boundary here—the extreme conditions of the Atacama Desert and the Andean highlands are effectively on a par with the interplanetary medium. This remarkably un-modern way of conceiving boundaries also permeates her work on an actual martian location, the Gusev crater (Cabrol was on the NASA team that oversaw the movements of the Spirit rover). The following excerpt from the same interview is particularly telling:

> We arrived at the limit between the hills and the plain, and there the rover took an image that will remain iconic for the Spirit mission. At first sight just a run-of-the-mill navigational photo in black and white, it is in fact a historic image. The front wheels of Spirit had reached the foothills, the Colombia hills, while the back wheels were still standing on the flat land it had traversed for the past six months. This means that the rover was positioned exactly between two worlds. A past world where Mars had conditions very similar to the Earth, and contemporary Mars as we know it today: cold and hostile. It was an incredibly powerful image, akin to standing on the Cretaceous–Tertiary boundary, akin to crossing the threshold between the world of dinosaurs and the world of mammals. (Cabrol 2022, translated from French by Praet)

In an obvious breach of modern habits of thought, ancient, aquatic Mars and the Cretaceous Earth are envisaged as independent planets in their own right: A geological boundary acquires the same status as the interplanetary medium. This unusual style of reasoning is by no means idiosyncratic. however. Consider what Juan Francisco Salazar (2018), an anthropologist who has done research with climate scientists on Antarctica, writes about ice cores. Such cores are cylinders of ice that are drawn out through drilling into the permafrost. He describes how ice cores function as time capsules, and as speculative devices that inform scientists about past and future Earths. Each core contains a sample of air bubbles, trapped from just a few centuries to several million years ago. Among other things, this bottled old air has allowed scientists to confirm that the concentration of carbon dioxide in the atmosphere is now almost forty percent higher than it was before the Industrial Revolution. Antarctic glaciers are thus conceptualized as archives of the terrestrial atmosphere. One could even say that they contain a record of ancient, miniature alien planets. As Salazar (2018, 40) articulates it,

> Ice cores are time probes. If a space probe is usually conceived of as an uncrewed spacecraft that travels through space to collect data that is sent back to Earth for analysis and interpretation, as is the case with several space probes launched in search of biosignatures of alien life, then ice cores can be conceived of as drilling into the deep time of extreme cryogenic environments in search of signatures of past climates. This is important when we consider that drilling ineluctably not only occurs through just ice, but also, as Jessica O'Reilly observes, potentially into "microbiospheres unique to the planet," . . . which may have been locked in time for several hundred thousand years.

What makes the parallel between ice core and outer space even more compelling is that O'Reilly (2016, 34), a colleague-anthropologist who also worked with scientists on Antarctica, specifically refers to Lake Vostok, a subglacial lake the water of which may have been isolated for some 15 million years or even longer. The lake is suspected to support a reduced-scale ecology of independently evolved lifeforms and is therefore also used as an astrobiological analogue site for ice-covered ocean moons such as Europa or Enceladus. In short, the upshot is that the Antarctic ice sheet is on a par with the interplanetary medium in this context. Against the modern Space Age convention, a geological boundary is treated as equivalent to a planetary boundary. It is a concrete example of what Messeri and Olson (Chapter 1, this volume) refer to as the pluralization of outer spaces. A further example relates to parrots.

Because of their remarkable longevity and their unmissable cleverness, parrots have been dubbed the humans of the avian world. The neuroscientist Claudio Mello has suggested that they are genetically as different from other birds as humans are from other primates (Klein 2018, Wirthlin et al. 2018). The literary historian Bruce Boehrer, author of *Parrot Culture: Our*

2500-Year Fascination with the World's Most Talkative Bird, has shown that parrot fanciers tend to lend their pets "a kind of honorary humanity" (2010: 46). The ecologist Kerry-Jayne Wilson (1990) calls the kea, an alpine parrot endemic to New Zealand, "a creature of curiosity" because of its manic playfulness, its taste for novelty, and its talent for mischief. The ornithologists Judy Diamond and Alan Bond (1999: 1) attribute these parrots "an extraordinary, alien intelligence." Upon closer inspection, keas possess a whole range of human- or at least primate-like features. They engage in allopreening—that is, they spend a lot of time preening the feathers of their fellow birds: "It resembles social grooming in apes or a handshake or hug between humans." (ibid. 66) They maintain a fairly strict social hierarchy through distinctive postures of the head feathers, which effectively function as facial expressions, and through a characteristic hunching display, which "looks a bit like a formal Japanese bow, giving the bird the air of a subservient samurai before the shogun" (ibid. 89). They even developed a form of bipedalism: With their relatively long legs, they can walk smoothly, and they are unexpectedly fast runners as well, albeit with an odd, skipping gait reminiscent of the hopping of lunar astronauts.

"They may sometimes behave like little green people," Bond and Diamond (2019, 16) note about parrots in general, "but their brains and their sensory organs are about as alien as those of crocodiles." Elsewhere in the same book (ibid. 6–7), they expand:

> It is inevitable that people should sense in these birds a familiar presence, a distorted mirror image of themselves: little, green, feathered people with a will of their own—clever, manipulative, capricious, rather like leprechauns. Through a fluke of nature, a peculiarly human intelligence seems to have been transplanted into the body of a bird. And the parallels are numerous: . . . Their social relationships enhance their ability to learn new tasks and discover new resources. And most impressively, like young humans, parrots from an early age acquire an ability to communicate vocally and to form long-term relationships with their friends and caregivers. . . . [Yet] their sensory systems operate strangely: They use their tongues to feel, rather than taste, the texture of food. They see a broad range of colors, but [like lizards] their two eyes provide independent views of the world.

This clear tendency to view parrots as alien humans rather than as familiar birds, as convergent aliens rather than as divergent life forms, as quasi-extraterrestrial beings with ultraviolet vision and the ability to use their tongue as a hand, is intriguingly prevalent in contemporary parrot research. Implicitly, and in breach of the modern convention, the species boundary between mammals and birds is reconfigured: It is no longer envisaged as a minor boundary between fellow living beings but, rather, as a major boundary, akin to the interplanetary medium, between two parallel planets. Similarly, New Zealand, the native habitat of the kea, has been reimagined as an extraterrestrial planetoid instead of a terrestrial island. The biogeographer Jared Diamond (1990, 3) famously stated that because of its astonishingly high rate of endemism, "New Zealand is as close as we will get to the opportunity to study life on another planet." In such biogeographical thought experiments, the Pacific Ocean is put on an equal footing with outer space. Along the same lines as Cabrol's separation between Earth-the-dinosaur-world and Earth-the-mammal-world, parrot researchers such as (Judy) Diamond and Bond arrive at a separation between a mammal-world and an alien but convergent bird-world that, besides parrots, also contains or used to contain giant moas, kiwis, and a whole range of flightless parrots, rails, and wrens—an otherworldly equivalent of the planet dominated by mammalian land dwellers that we humans are most familiar with. It is another telltale instance of the pluralization of outer spaces.

What the examples of planetary geology (Cabrol), the anthropology of science (Salazar), and parrot research (Bond and Diamond) show is that "actual" space exploration in orbit and beyond is only one—albeit curiously fetishized—way among many possible ways of engaging with a whole gamut of unsuspected "outers." By and large, the latter have remained under the radar. Scholars in outer space studies are beginning to realize that cutting-edge exploration is not necessarily rocket-based exploration: Used well, a good piece of drilling equipment or a decent pair of binoculars may be as pathbreaking as the most sophisticated, artificial intelligence–enhanced space telescope. A judiciously executed investigation of extreme, ultraviolet-radiated lakes in the high Andes may be as insightful as a year's worth of astronautic experimentation on the International Space Station (ISS), with its annual operations cost of approximately $3 billion for the United States alone (see Paola Castaño [2021)] for an original take on the question of the scientific value of the ISS). Exoplanet hunter Martin Dominik, interviewed by Praet in 2024, offers a further illustration: The detection of OGLE-2005-BLG-390Lb, a so-called Super-Earth, which he masterminded and which is now recognized as one of the great astronomical discoveries of the early

2000s, happened by means of a network of supposedly "creaky, old, ground-based telescopes" such as the modest Danish 1.54 meter telescope at La Silla (Chile). Shortly before the momentous detection, officials from the European Southern Observatory were about to decommission the instrument, which dates back to the 1970s—it was only deemed good for the scrap heap.

The Inescapable Provinciality of Space Exploration

In short, there is a growing awareness that the most innovative advances often come from unexpected corners and from the type of unorthodox research I just described, rather than from the overhyped schemes of Silicon Valley loudmouths. In that sense, classic slogans such as "reaching for the stars" also betray a certain kind of narrow-mindedness, and a distinctive lack of imagination. The acclaimed light artist James Turrell, who closely engaged with the aerospace industry in his early career, once quipped that modern space travel is less a capability than a deficit—the inflated outgrowth of our incapacity to see the cosmos in a rock. Contrasting American and Japanese attitudes, Turrell, an American, perceptively complained, "When we want to go into the universe, we can't look at a rock, like the Japanese. We have to actually go to the moon. We're so literal" (Waterhouse 2022, 748) This compulsive belief in literalness is a signature of Space Age modernity, but it is also a marker of a mostly unacknowledged kind of conceptual provinciality. The universe, as Turrell's entire oeuvre testifies, does not *have* to be thought in this literal manner. Otherwhere ethnography can be pictured as an offshoot of this reaction against the all-pervading cult of literalness. In outer space studies, there is indeed a growing awareness that space enterprises must be understood as inherently provincial undertakings. This move to "de-fetishize" the interplanetary medium and to "provincialize" contemporary space exploration is not an argument for unabashed relativism, however.

Otherwhere ethnographers have no interest in rekindling the tired old quarrel between universalists and relativists. None of the contributors to this volume seek to enlarge the supposed rift between the natural sciences and the social sciences. Otherwhere ethnographers may have misgivings about scientism, but they are emphatically not anti-science. They rather seek to understand the space sciences in a more nuanced way. Our volume reveals that natural scientific disciplines such as astrobiology, planetary geology, and astronomy are more interesting and internally more diverse than modern accounts of them typically acknowledge. As Zara Mirmalek notes in Chapter 6 on the multidisciplinary BASALT program, in which scientific teams cooperate to investigate volcanic terrains on Earth as analogs of Mars, although different scientific communities are assumed to work in harmony, a range of situated differences (e.g., institutional goals and public accountability) also make them "incongruently distinct." Astrobiology, Praet suggests in Chapter 3, is not of a piece: It encompasses incompatible metaphysical styles, with long and equally venerable genealogies. Its roots are, in the words of Campos, "resolutely rhizomatic." What transpires, if one collates the latest findings of anthropologists, historians, and sociologists of science, is neither the inexorable advance of a front of objectivity nor the gradual unveiling of one singular truth about the universe. Instead, what emerges is a mosaic of objectivities and a diversity of truths-in-motion. The task at hand for otherwhere ethnographers is to elucidate these "cultures of the scientific knowledge production," as Mirmalek phrases it in Chapter 6. As the modern belief in one purified Nature recedes, they are grappling with a situation in which a multitude of subtly distinct natures coexist.

The trouble with scientism and the concomitant beliefs in literalness and in nature is ultimately that it is premised on an utterly unrealistic notion of science and on an astonishingly naive understanding of how the most creative space researchers actually operate. Emblematic figures such as Bezos and Musk like to picture themselves as champions of science, but their very conception of science as a smoothly unified enterprise geared toward progressive enlightenment and cumulative prosperity for humanity as a whole is—most, if not all. authors of this volume fear—both terribly outdated and seriously misguided. As fetishists of the interplanetary medium they are positively quaint. One might indeed view them and their numerous emulators around the globe as relicts from a bygone period—Space Age modernity. What this book offers is a more heedful kind of optimism and a less conformist idea of exploration: Otherwhere ethnography reveals that space exploration, in theory, does not need to be imagined along the lines of the neocolonial templates that space billionaires are so strongly attracted to. De-fetishizing the interplanetary medium and pluralizing exploration is not only thinkable but also, the chapters in this book show, already happening right, left, and center. The vocation of otherwhere ethnography is to be the antenna that captures this momentous conceptual mutation. In Chapter 7, Aaron Parkhurst and David Jeevendrampillai describe how extraterrestrial restaurants are being envisioned. To expand on their restaurant idea, instead of a hot dog caravan with only three types of bland sausage (SpaceX, Blue Origin, and Virgin Galactic), one might want to picture a quirky street food stall with an unexpectedly varied menu, including zero-gravity zucchinis and other vegetarian options.

PART I

PROVINCIALIZING THE SPACE AGE

Historical and Anthropological Perspectives on the Exploration of Outer Space

1

Spaced Out

Bringing Outer Spaces into Anthropological Conversations

Lisa Messeri and Valerie Olson

Introduction

Outer space is the anglophone designation for space beyond Earth's atmosphere. Scholars interested in the social dimensions of that space initially conceptualized it as an historical and geopolitical category, doing so in response to post–World War II U.S. investment in the physical sciences (including astronomy) and even more significantly the Cold War space race. Conceptualizing outer space as an *ethnographic* category of study has been a more recent development. To propose an anthropology of outer space challenges its relegation as a singular monolithic space "outside" the discipline, generally considered off limits to fieldwork (Olson 2023) and only contingently inhabited by elite subjects. Nevertheless, even as early as the 1970s, anthropological experiments began to bring outer spatial things, people, places, and processes into ethnographic projects. Among the results of these ongoing projects is that "outer space" is now acknowledged as a plural space—"outer spaces"—for social groups as well as analysts. This means that outer space is not an undifferentiated region only relevant to geopolitical elites enacting colonial projects but, rather, there are multiple ways of knowing, historizing, and relating to infinite spaces and places beyond Earth's atmosphere.

Today there are proliferating anthropological inclusions of these outer spaces in a variety of research projects, not only those exclusively oriented toward the Western cosmos. But questions remain about why and how anthropologists should study these spaces. In this chapter, we address what happens when anthropologists include outer spaces within anthropological

Lisa Messeri and Valerie Olson, *Spaced Out*. In: *Otherwhere Ethnography*. Edited by: Istvan Praet and Perig Pitrou, Oxford University Press. © Oxford University Press (2025). DOI: 10.1093/9780197790885.003.0002 space. We do so by surveying studies in which scholars include human relationships to outer spaces by doing observations, surveys, interviews, and material culture analyses. We show how these inclusions end up confronting conceptual binaries that constitute anthropological conceptualizations of "space" as a general category. The binaries we are primarily interested in here are those that ethnographers of outer spaces simply cannot maintain as they do their work.

When it comes to understanding social spaces in general, most anthropologists use—and criticize—categories that differentiate physical and conceptual spatialities. These can include binary dyads such as here/there, known/unknown, local/global. As Marilyn Strathern (2004) noted, such categories are entrenched in Western, specifically anglophone, spatial and scalar schemas of relational order. These schemas have long influenced social scientific distinctions of entities, things, and spaces in their research projects. Anthropologists may critique these categorical schemas today, but they still work with spatial distinctions in order to form their inquiries, shape analyzes, and facilitate disciplinary communication.

One of anthropology's least-critiqued binary divides is that between "earthly space" and "outer space." In this divide, nonterrestrial spaces can be premarked as disconnected *other* spaces. They become hyper-outer, existing outside of anthropology's many worldly and otherworldly spaces (Zhan 2009). Such outer spaces and their things are assumed to be *extra*, in the sense of supplementary or excessive, to everyday life on earthly ground. As a result, outer spatial materialities, immaterialities, worldlinesses, epistemological statuses, or ontological dimensions are often treated as extraneous to anthropology's central concerns or processes. Some scholars even suggest the need to refocus *downward* or earthward (see Latour 2018) so as not to be distracted by the (singularly imagined) nonterrestrial, further implying that there are no politics involved in de facto Western terrestrial–centrism or in the binaristic separation of terrestrial spaces from outer spaces and things. Not only have celestial objects long figured in non-Western social and political organization but also scientific studies of the solar system have been foundational to Earth systems science and politics (Olson and Messeri 2015). Focusing downward risks missing these connections.

We structure our survey of anthropological projects in outer spaces by addressing their challenge to three binary/boundary schemas: *material/immaterial*, *terrestrial/extraterrestrial*, and *epistemology/ontology*. In particular, we call attention to how challenges to these schema have productively intensified through new topical foci in studies of outer spaces.

We show how ethnographers of outer spaces are increasingly challenging the assumptions that mark space as extra-anthropological in physical and conceptual terms, and in doing so, they challenge assumptions about what counts as anthropological space. These scholars reveal how spatial prefixes, such as outer and extra, naturalize overt or covert divides of disciplinary attention or practice that make outer space anthropology seem inherently inhuman or unearthly. In addition, they refuse to segregate outer spaces as zones in a separate non-living domain beyond anthropology's life-centered bounds. Contemporary scholarly voices are also calling for ways to break free of Western liberal, racialized, and settler colonial scientific conceptualizations of outer spaces that erase social experiences of contiguous life-spaces that are not limited to scientific planetary boundaries. Ethnographers of outer spaces include them as zones that anthropology can and must go to track the scopes and scales of power, race, gender, sovereignty, embodiment, and futurity (Olson 2023). We hold that this is relevant for all scholars calling hegemonic spatial conceptualizations into question during the project design, fieldwork, and analysis phases of their work (Peterson and Olson 2024).

Questioning the Assumed Spatial Dimensions of Material/Immaterial Binaries

For many social scientists, outer spaces seem to be inherently disconnected, unmoored, and ultimately inconsequential to social life. In presumptively *not mattering*, outer spaces get mistaken for not having material presence on Earth. In essence, outer spaces are rendered immaterial. Space studies take exactly the opposite approach. They not only make cases for including outer spaces in social science but also call attention to the hidden earthly/unearthly binary that can haunt material/immaterial divides. In this section, we show how outer space studies highlight the hidden spatial division that associates earthly spaces with materiality and social significance and outer spaces with immateriality and social insignificance.

Ironically, it was archeology, anthropology's most materially focused subdiscipline, that engaged outer space in a way that rendered the material/immaterial binary untenable in the 20th century. In the wake of this work, 21st-century space studies scholars are reframing outer space as an open space for (re)imagining human pasts, presents, and futures. This outer space can be a space of immaterialized imagination, but it is also a space of materializing speculative connections between earthly and unearthly spaces and things.

Material Inscriptions of Imaginative Space

Archaeology is an anthropological subdiscipline in which outer space is not "outer" in cultural historical terms, either conceptually or materially. Archaeologists readily include suns, moons, stars, planets, and meteorites in their analyses of key artifactual symbols. Outer space and its objects are treated as imaginative forms when people mobilize them as symbols to shape social practices and organization, such as through astrological prediction or when they signify transcendental dwelling spaces such as heaven. While the archaeological engagement with extraterrestrial things as symbolic thus seems to collapse the earth/space boundary, it also creates a boundary between the imaginative and material (particularly in the sense of shaping terrestrial practices).

Outer spaces and objects are categorized as material cultural forces when they act as guides for agriculture and migration. A notable group of archaeologists advance this work by treating spaceflight sites and materials as contemporary migratory forms. One example of late 20th-century archaeology/anthropology "crossover" work by anthropologists of star-based ocean navigation practices among Pacific island societies theorized contemporary spaceflight as a continuation of a "human" migratory imperative (Finney and Jones 1985). In the early 21st century, archaeologist Alice Gorman began work with colleagues to treat and preserve spaceflight machines and launch and landing sites as archaeologically significant materials of heritage and colonization (Gorman 2005a). Working in this mode, "space archaeologists" produce critical studies of spaceflight material culture in situ—on Earth, on other planetary ground, and in orbit—as forms that shape the interconnected political and social worlds of those who build them and are impacted by them, including Indigenous societies (Gorman 2005b; Gorman and O'Leary 2013; Buchli 2020).

Recently, anthropologists of religion have, like archeologists and material studies scholars, offered outer spaces as a productive way of thinking about the relationship between the material and the immaterial. Deanna Weibel (2020), for example, in writing about the overview effect—an experience of awe often described by astronauts—draws on Timothy Morton's (2013) theory of hyperobjects to ground the religious experiences induced by human spaceflight in the materiality of the cosmos (see also Weibel and Swanson 2006). Outer spaces are both imagined and embodied, and although attempts to launch grounded inquiries—such as archeology and religious studies—into space might be treated with skepticism, it is this dual nature of outer spaces that facilitates questioning the distinction between the material and immaterial.

From Imaginative Spaces to Speculative Spaces

While wrestling with the seemingly extreme immaterialities of cosmic imaginaries—the far-off utopian, inhuman alien, and unearthly speculative—scholars have repeatedly traced the materiality of these imaginations to open new research trajectories into the extensively trans-spatial and extra-planetary dimensions of everyday social experience.

Some of the first outer space studies focused productively on the utopian, transcendental, and ritualistic dimensions of collective enterprises that move matter, including bodies, beyond Earth's atmosphere. In 1978, William Sims Bainbridge conducted a study which holds that U.S. human spaceflight remains a militarily and politically unsupportable goal that continues because of social advocacy in favor of its utopian transcendental value. With collaborators, he continues to collect and analyze survey data on spaceflight's shifting ideological support and meanings. In the 1970s, anthropologists also delicately dipped a toe into extraterrestrial waters by becoming involved in conversations about alien contact (Maruyama and Harkins 1975; Collins 2003; Dick 2020). After a notable two-decade gap in outer space–focused qualitative research, Debbora Battaglia (2006) edited a volume on the extraterrestrially alien as a marker of quasi-scientific and quasi-transcendental otherness that imbues figures in North America with uncanny social power. The alien continues to be a culturally revealing figure. Susan Lepselter's (2016) multidecade study and reflection on alien abduction support groups searches for—along with her interlocutors—the material traces of "unseen things." The alien is not allowed to be exclusively imaginative and, similar to the thrust of Battaglia's volume, is always already shaping earthly materiality (see also Dean 1998, Penley 1997).

Likewise, in the social sciences more broadly, speculation has recently come to *matter* as a mode of engaging with unthinkable presents and futures (cf. Benjamin 2016; Haraway 2016; Anderson et al. 2018). Speculation is no longer a synonym of imagining but, instead, a method for making material what people, including social scientists, wish to manifest. In anthropologies of outer spaces, it becomes apparent that speculations about connections between terrestrial and nonterrestrial spaces and things can become vitally material on Earth. Speculation is being employed as an alternative way of approaching complex, planetary crises—problems for which the ways of knowing currently deployed have proven insufficient (Bratton 2019). The expansive material differences of outer spaces are precisely what make them productive for approaching such challenges anew. Just as aerospace engineers and new space entrepreneurs tether outer spatial imaginative impulses to material ends, inscribing pasts and futures into space infrastructures (Scharmen 2019b; Tutton 2018, 2021), scholarly speculative projects have similarly been animated by an unbounded cosmos. Crafting speculative texts implicitly acknowledges the need for radically other ways of relating to outer spaces (see Shorter and TallBear 2021; see also Chapter 4, this volume), a point we return to in the third section of this chapter.

In outer space studies, outer spaces are folded into a conjoined material and speculative space that pushes against an assumed association of Earth with materiality and outer space with immateriality. In this work, earthly/unearthly boundaries are not taken for granted, calling attention to the problem of associating Earth with realness and the unearthly with exception and excess.

Terrestrial/Extraterrestrial Crossings

In addition to revealing the hidden earthly/unearthly boundaries of the material/immaterial binary, space studies also call attention to the anthropological importance of the terrestrial/extraterrestrial binary. Space studies scholars do not simply deny this spatial division. Instead, they pay attention to how it is socially constructed through shoring up and breaking down discursive and physical boundaries. Much of that work calls new attention to a key concept that conjoins both terms: *territory*. Space studies scholars do not always assume that outer spaces are *extra*territorial in the geopolitical sense of being outside the law. Rather, they often attend to how outer spaces are quasi-territorial, in that it is in the process of being selectively territorialized and extraterritorialized, and that geopolitics itself is not bounded at the "geo."

As space studies scholars call attention to the social production of terrestrial/extraterrestrial divides and crossings, they call attention to the tacit persistence of this divide within social analysis. They investigate outer spaces' quasi-territorality across several dimensions, including imperial infrastructure building, the production of embodied subjects, and the inscription of trans-spatial territorial spaces and powers. This pushes investigations of territory and territorialization into new forms and spaces, from the extraterritorial activities of space flight and the space sciences to the scientific designation of "outer space" as a terrestrial-containing environment or ecosystem.

Extra-territories and Their Embodied Subjects

One of the earliest and persistent focuses of space studies is how terrestrial boundary crossing machines—rockets, telescopes, satellites, robotic craft—produce new forms of extraterritorialized knowledge production and claims. In this register, scholars examine the extraterrestrial dimensions of nation-building, colonization, and imperialism. Peter Redfield's (2000, 2002) innovative work on colonial spaceport building in French Guiana frames it as a powerful imperial spatial technology (see also Mitchell 2017). In his multivolume history of the rocket as a Soviet national unity-building machine, Asif Siddiqi (2000, 2010) argues that outer space was a crucial extension of Cold War industrial territory. More recently, studies of New Space industrialists have argued that colonial dreams persist and are enhanced by capitalism's desires (Genovese 2017; Tutton 2018, 2021; Sammler and Lynch 2021a). These newest claimants of territory beyond Earth represent an economic sector outside the big government contractors of the mid-20th century and view outer space as an "exit strategy" for terrestrial economic emplacement (Valentine 2012). Outer spaces offer new extraterritorial capitalist contexts, potentially outside legal and social constrictions on extraction and production (Klinger 2018).

Rockets—carrying both people and cargo—in these works are off-world material extensions of a foundational imperial technology: the telescope (Schaffer 2007; Shorter and TallBear 2021). In his investigations of Western astronomy as a cosmological practice, Goetz Hoeppe (2012) shows how astronomers territorialize cosmic images and data in their observatories and computers to validate the truth-value of the work. Although space scientists—in distinction to New Space capitalists—might deny the territorialization of their work (after all, the power of science in contemporary society comes from a denial of its politics, even if this outward performance might contrast with how a scientist individually or personally understands their work), they nonetheless produce terrestrial and extraterrestrial *places* in tandem. Ethnographers working with such scientific experts note that although some of these groups purposely strive to keep material places distinct from the imaginative places, they end up facilitating breakdowns between what counts as concrete national space and place and the pursuit of projects aimed at extraterritorial planetary or interstellar places. Lisa Messeri's (2016) ethnography traced these spatial overlaps from analog projects (see also Olson 2018; Black 2018) to the territorialization of near and far planets as scientists marked extraterrestrial worlds as places. In a variety of scientific endeavors, the distinction between Earth and not-Earth blurred in pursuit of creating knowledge about the cosmos (Messeri 2017).

In addition to showing how spaceflight technologies and the space sciences both make and cross the terrestrial/extraterrestrial binary in order to territorialize power, authority, and knowledge, scholars began to look at the bodies working on and at that intersection. Instead of assuming that people working in space military/industrial sites would have bodies unfazed by outer space's nonterrestrial conditions, ethnographers began to ask what kinds of bodies space-territorializing societies want.

One outcome of space studies is new understandings of contemporary industrial embodiment as more than terrestrial. In particular, some studies focus on how workers flown into outer spaces are doing more than extending geographic power into space. They maintain the exclusive power of transgressing the terrestrial/extraterrestrial divide in order to territorialize new forms of embodied occupational power. In her study of government astronaut embodiment in the United States, Valerie Olson (2018) argues that making outer space into normal territory requires recategorizing what counts as a medically "normal" U.S. body. Astronauts in space exhibit impairments considered medically abnormal on Earth, but those conditions are officially recategorized as "outer space normal." This maintains the social status of astronaut elite embodiment, while making outer spaces into normalized extensions of U.S. territory. In his study of the making of astronauts as government workers, Matthew Hersch (2012) shows how midcentury astronauts, as socially exceptional but institutionally controlled White males, became specially protected laborers who extended governmental organizational structures beyond Earth's surface. In her ethnographic work among U.S. space program teams operating robotic craft off Earth, Janet Vertesi (2015) shows how workers become sensorily enmeshed with their machines in ways that are not typical of other earthly workers. They must "be" rather than simply operate those craft.

Extended Territories, Environments, and Ecosystems

By examining how outer spaces become politically and economically linked to earthly places, outer space studies call attention to the terrestrial/extraterrestrial binary embedded in the concept of "geography." The concept of geo-graphy and its attendant geos, such as geopolitics, naturalizes a political sphere divided between terrestrial places and extraterrestrial spaces. However, in many of the works referenced above, outer spaces are not just linked to terrestrial places; they become places in their own right. And emerging work by scholars examining the politics of including outer places and things in national and corporate enterprises, from rocket launches to asteroid mining, shows that today's geopolitics and the "ground" of capitalism are post-geographic and post-geological (Olson 2012, 2013; Valentine 2012; Jacka 2018; Kinchy et al. 2018). In this space of extensions from Earth, the political sphere is helio-scale, including solar, lunar, asteroidal, and interplanetary spaces.

As scholars began to analyze outer space's manifestations as quasi-territorialized space and place, they also turned to what it was becoming other than "space," namely an environment and ecosystem. For example, emerging work on the environmental geopolitics of outer space shows how national and corporate engagements with outer spaces extend and intensify the impacts of environmental power, racism, degradation, global divisions, and sacrifice (Olson 2018; Klinger 2021). Other scholars examine how extraterrestrial industrial and military activities were vital to the production of new transterrestrial forms of broad spatial relations knowledge and governance (Deloughrey 2013). Sheila Jasanoff and Marybeth Long Martello (2004) deploy a visual ethnographic analysis to the production of environmental images from low Earth orbit, showing how they were presented by spacefaring nations as neutral depictions of "the human environment" but were actually specific bids to create a geopolitical high ground for countries empowered to define "the planetary environment" as a political space. Low Earth orbit, a formerly outer space, is a zone of geopolitical surveillance as well as a site for the extension of technological property and industrial pollution (Rand 2019).

Although outer space studies seem spatially focused, in the next section we show how those studies call attention to the domains of knowledge, being, and becoming. In these studies, outer spaces are existential spaces.

There Are No 'Ologies in Outer Spaces

Space studies work does more than problematize dominant Western spatial binaries and boundaries; it is also problematizing dominant *existential* categorizations and boundaries. To do this, space studies scholars are calling into question a set of cosmological dividing practices. One division in question is life versus nonlife, the other is epistemology versus ontology. Space studies scholars show that in outer spaces, basic Western cosmological assumptions about the separability of life and nonlife and the distinction of knowing and being cannot be maintained as routine givens in anthropological inquiry. Scholars investigating extensions of Western science into outer spaces expose how "biological life" is not lively enough to be cosmologically valid.

As a result, ethnographers who include outer spaces find themselves working within a trans-spatial zone of liveliness that undoes the terrestrial/extraterrestrial and life/nonlife divides. In addition, scholars in the social sciences and humanities are questioning the "ology" divisions that fuel attention to ontology over epistemology, and either one over cosmology. Showing how projects to "know" or "be" in space reinscribe colonial territorializations and marginalization of otherwise existential experiences and authorities, emerging scholarship is powerfully de-othering outer space studies without resorting to calls for a "cosmological turn."

Liveliness as a Shifting Space

While the work we surveyed in the previous sections was pragmatically oriented to life in relation to spaces and places, space studies scholars also engage with the problem of life as an existential condition, particularly with the existential contingencies of imagined or tangible unearthly life. Engagements with extraterrestrial or alien life as a discursive or material formation are always fraught, for both the sciences and the social scientists. Not only is the validity or plausibility of the whole inquiry anthropologically questionable but also the question of how to productively study "life as we don't know it" remains an open question. As a result, the hidden "other" and "outer" forms that animate the concept of "life itself" in anthropology come to the fore: the nonbiologically lively and the nonliving.

When faced with the possibility of otherly animate dimensions of outer spaces, neither space science nor anthropology can maintain a strong life versus nonlife boundary. Scientists looking for life elsewhere than Earth are confronted with working with imaginative, unrecognizable, and counter-life, and anthropologists who study the conceptualization of life, such as in the biological sciences, find themselves working with biologists without life (as we know it) to work with.

Scholars who study exobiologists document the terrestrial/extraterrestrial boundary tensions that emerge when scientists attempt to theorize life on terrestrial terms specifically or on planetary terms in general. In Claire Webb's (2021) historical exploration of the exobiological imagination, she notes the mid-century origin of a "planet-as-island imaginary." Terrestrial islands are imagined both as enclosed ecosystems and places to expand to and from. In imagining planets as islands, Webb argues that this tension between enclosure and expansion was propelled into the cosmos. The tensions between searching for life versus searching for liveliness simultaneously expand what might exist in the universe and contract or enclose normative notions of life.

Anthropological attention to the problem of what counts as life on or off Earth calls attention to the continued situated binarism of Western notions of "consciousness/unconsciousness," "intelligence/untellegence," or "life/nonlife." Anthropologists have focused on SETI (the Search for Extraterrestrial Intelligence) as a site where these categories are being actively addressed (Denning and Berea 2019; Wright and Oman-Reagan 2018; see also Capova 2013; Traphagan 2016). SETI's history marks it as entrenched in frontier and colonial mythos and thus unlikely to be able to lead a radical rethinking of liveliness, even if individuals within this heterogeneous project critically intervene with SETI's core tenets to maintain such aspirations.

Indigenous scholars addressing SETI's project call attention to the colonially framed conceptual frameworks that define private and governmental "searches for life" and "consciousness" at large (Charbonneau 2021; Lempert 2021; Shorter and TallBear 2021). In their analyses, these scholars show that settler colonial science not only reinforces projects of discovery and enforced contact but also reinscribes divisions between human-as-Western-man and life-as-Western-being and every other form that does not fit those criteria. In this model, every other form defaults to non-being-as-thing (Atalay et al. 2021). Transformed attention to outer spaces, then, requires that anthropologists question and abandon distinctions between life-as-being and nonlife-as-thing on and off Earth.

To take astrobiology seriously as an object of social scientific study necessarily transforms thinking about life on the level of the organism, which decenters a basic anthropological concern with life as an embodied form. Scholars working on these topics note how the search for intelligence and the search for traces of biological life require treating nonliving things, such as signals or geology, as if they are either existentially equivalent or unequivalent to life. Astrobiologists are necessarily forced to make life strangely nonbiological, and thus a productive resource for unsettling given terrestrially centered assumptions and distinctions between life and not-life (Salazar 2017a; Helmreich 2006, 2009). Istvan Praet (2017) has further demonstrated that astrobiology helpfully unsettles not only categories of life but also disciplinary boundaries (see also Marcheselli 2019).

Space studies scholars also show how exobiological thought can push the bounds of anthropological thought in terms of liveliness not only as a characteristic of an organism but also as a characteristic of a planet. Messeri noticed a subtle shift between when she was conducting her ethnographic research with exoplanet astronomers in 2009 and 2010 and how scientists have more recently talked about the ambition of their field. Whereas it used to be off limits to connect the search for an Earth-like exoplanet to a desire to find life, this has become less taboo. When exoplanet astronomers discuss signatures of life, they are not imagining a single being beyond the data but, rather, a planet as a whole that is lively. The search for another Earth is the search for a living planet. As imaginations of other Earths proliferate, so does the imagination of Earth-as-Other (Messeri 2017). Helmreich (2012) plays with this tension, naming it "extraterrestrial relativism." For scientists searching for other Earths, our planet becomes something relativized, othered, and compared. As Helmreich notes, the imagination of Earth as other—as Spaceship Earth or Gaia—relativizes Earth in that "it becomes not only one planet among others, but also a planet that points to and even contains its others" (1133). The search for an Earth-like exoplanet, in contrast, leads to "reimaginings of Earth as other than it is—a kind of speculative extraterrestrial relativism, bent back to respin 'Earth'" (1132).

In summary, anthropological critiques of scientific life/nonlife boundary-keeping can be enhanced by focusing on what happens when those boundaries extend off-Earth. These binaries can become strengthened, as in the case of SETI and some astrobiological projects, or, in the case of other astrobiological projects, destabilized. As a result, scientists and anthropologists alike recognize that "liveliness" is a shifting category. Thinking this category beyond the solar system changes the scope, scale, and, subsequently, the questions that can be asked of and about forms of existence.

Refusing Cosmos and Cosmology

In addition to confronting the existential binary of life/nonlife, space studies scholars also confront the epistemological boundary between positivist and otherwise forms of knowing. One aspect of this boundary includes the injunction that scientific cosmologies are epistemologically and ontologically true, while other cosmologies are merely metaphorical or *immaterial*. Statements such as this are haunted by the specter of Western divisions between knowing, being, reality, unreality, space, and time. Even as anthropology pursues corrective scholarly actions through "ological turns"—epistemological, ontological, and even cosmological—space studies scholars engage with ways of knowing and experiencing the existence that cannot be neatly categorized. In recent work (often coming from beyond anthropology) focused on situated outer spatial knowledges that run counter to or resist the perpetuation of Western knowledge-making projects, the liberal formulations of cosmos and cosmology decouple from people's actual and speculative relationships with spaces outside of Earth's atmosphere.

Space studies scholars highlight pre-existing forms of terrestrially unbounded spatial knowing that have historically been dismissed in a Western, White academic context. Folklorist Jane Young (1987) documented how different communities of First Nations and Indigenous North Americans reacted to the Apollo moon landing, largely unimpressed due to a cosmology that does not conceive of disconnected, empty space between the Earth and moon. This analysis by a non-Indigenous scholar still contained elements of exoticism even while challenging the idea of a singular "cowboy" cosmology. More recently, Indigenous scholars have transformed conversations in the social studies of outer space by centering on land and relations as concepts necessary to understand the politics of the contemporary space sciences. Put bluntly, those

> who write from the vantage point of Indigenous studies are not so afraid of the unknown extraterrestrials, the vastness of space, or the farthest depths of the galaxy and beyond. Instead, we fear that various space exploration initiatives are reembodying the attitudes and practices of terrestrial explorers in the past. (Shorter and TallBear 2021, 2)

In new space studies projects focused on land and relations, imperial and colonial telescopic cosmology-making are vehicles to expose the provincial Western "cosmos" concept that hides within the universalistic concept of "cosmology." Kānaka Maoli have protested the construction of the Thirty Meter Telescope (TMT) on Maunakea since 2014, citing the astronomical community's long disregard for Indigenous religious and land rites (Casumbal-Salazar 2017; Goodyear-Ka'ōpua 2017; Maile 2021; Sammler and Lynch 2021b). Hi'ilei Julia Hobart (2019) draws out how science and capitalism have worked hand in hand to transform Maunakea into an empty space that can serve as a gateway for the galaxy, depriving the Mauna of unbounded experiences of animacy. Deondre Smiles (2020), building on the settler colonial critique of the TMT, suggests that native conceptions of space offer another way of enacting human being in the cosmos (see also Kite 2021; Lempert 2021). The importance of land extends beyond specific telescope projects, as Fantasia Painter (2021) demonstrates in her exploration of how Indian land shaped the U.S. Federal Bureau of Investigation's UFO investigations in the 1970s. Collectively, this scholarship builds on broader anthropological trends that center the futurity of Indigenous scholars, communities, and cosmologies (see also Harjo 2019).

In addition to centering Native scholars and cosmologies, the anthropology of outer space must attend to the wide range of situated knowledges that expose and transform dominant definitions of outer-space-as-cosmos. Astronomer Chanda Prescod-Weinstein (2021), drawing on her experience as a Black cosmologists, has argued that the inequities prevalent in the institutions that produce scientific knowledge and knowers become embedded in the knowledge produced. Scholars in science and technology studies would be quick to co-sign this observation, but Prescod-Weinstein further demonstrates that a Black feminist physics opens up new avenues for studying the universe (see also Prescod-Weinstein 2020).

Conclusion: Outer Spaces Are Anthropological Spaces

What happens when outer spaces are not marked as excessive or naturally othered spaces, places, environments, or cosmoses in anthropology? We have reviewed the ways in which outer space studies open anthropology's space-at-large and opens questions about the wholesale spatialization of other categorical binaries. As a result of these works, binaries such as the imaginative/material, terrestrial/extraterrestrial, and epistemology/cosmology or ontology/cosmology can no longer be unquestioningly pushed into an outer off-world. As such, anthropology's primary categorical concerns, including race, gender, and ability, can be examined without bounds. Projects in outer spaces are therefore not anthropological excessive or extraneous but, rather, are necessary spatial extensions of questions and problems that connect all critically engaged anthropologists.

We consider these projects transformative in that they refuse to reproduce given divisions in social as well as cosmic thought. As "outer space" has become a place ever more entrenched in conquest and capitalism—with the billionaire space race serving as a stark reminder—scholars subvert this singular imaginary by attending to the vastness of outer spaces, which have room for infinite projects, critiques, and futures. Although the projects discussed in this chapter do not fully escape the gravity of binary thinking, they contain aspirations for otherwise unbounded ways of doing anthropology. The appeal of having sites for unbounded thinking has only grown as the bounds of the earthly politics and practicalities have felt more urgent and constraining. This is not to say that outer space studies are fleeing Earth. In fact, as we have shown, it is quite the opposite. In an effort to find new modes of terrestrial thinking and being, experiments of spatial and temporal boundary breaking are being performed in outer spaces but always in relation to experiences in the grounded present.

2

Blue Vegetation on the Red Planet

Soviet Astrobotany and Earthly Analogs for Life on Mars

Luis A. Campos

Tikhov and the Making of Astrobotany

The earthly roots of astrobiology and its many varied efforts to explore the history, prospects, and limits of life in the universe are resolutely rhizomatic. One prominent though often overlooked shoot is the field of "astrobotany," developed by the Soviet astronomer Gavriil Adrianovich Tikhov (1875–1960). From his considerations of what the Earth must look like to an observer on the moon, and his studies of Earth's reflected light leading him to conclude that the Earth must appear from a distance to be "a disc of a pale blue color,"[1] Tikhov developed the transplanetary uses of astrospectrophotometry more generally to understand life elsewhere in the solar system. While American scientists such as Joshua Lederberg (who coined the term "exobiology") and Carl Sagan (who had the works of Tikhov translated and would later popularize the idea of Earth as a "pale blue dot") advocated for new approaches to the search for life beyond Earth that inspired generations of scientists,[2] Tikhov's methods, context, and stated commitment to Soviet principles of dialectical materialism for advancing science offer a different sort of rhizomic red thread in the history of astrobiology and the red planet. Although forgotten and unfamiliar to many practitioners of today's astrobiology, the case of Soviet astrobotany provides a fascinating example of how Earth and Mars have long been tied together in more ways than one, and it shows how the history of terrestrial politics is an inescapable feature of our understandings of extraterrestrial elsewheres.

It was in his autobiography, *Sixty Years at the Telescope*,[3] that Tikhov revealed where his passion for studying the heavens came from. Although

Luis A. Campos, *Blue Vegetation on the Red Planet*. In: *Otherwhere Ethnography*. Edited by: Istvan Praet and Perig Pitrou, Oxford University Press. © Oxford University Press (2025). DOI: 10.1093/9780197790885.003.0003 born near Minsk, it was as a young man in the Simferopol Public Library in the spring of 1892 that Tikhov first read Camille Flammarion's *The History of the Heavens* in Russian translation and *Popular Astronomy* in French, and decided to become an astronomer:

> Avec avidité je me procurai des livres d'astronomie sans cesse renouvelés. Chaque lecture me dévoilait l'une après l'autre les énigmes des astres. L'Univers gagnait en étendue et en mystère, et mon ravissement ne connaissait pas de bornes devant son immense variété et sa beauté inégalable.[4]

As a teenager, he built his own observatory while he was also taught by his grandmother about flowers. "At that time," he later recalled, "I hadn't supposed that astronomy and botany would merge for me in many years, and at the junction of these sciences astrobotany would be created—the science connecting the distant planets and plant life."[5]

In 1895, he also became very interested in botany and read several books, including the remarkable *Plant Life* by Kliment A. Timiryazev, who had "considered photosynthesis . . . as the interrelatedness of terrestrial and cosmic processes," giving plants a cosmic role to play:

> La feuille verte, ou plutôt le microscopique grain vert de la chlorophylle, est un foyer, un point de l'Espace universel vers lequel, d'une part, converge l'énergie du Soleil et à partir duquel, de l'autre, prennent leur naissance toutes les manifestations de la Vie sur Terre. La plante est un intermédiaire entre le ciel et la terre. Voilà le veritable Prométhée qui a dérobé la feu du ciel . . . le principe de base de l'organisme vegetal dépend effectivement de la lumière, il devient evident que l'on doit rechercher la principale particularité de la plante dans ses propriétés optiques également.

Soon after he married in 1898, he traveled to France and, while enrolling at the Sorbonne for some classes, primarily studied at the Meudon Observatory, which had made its fame through solar photography under its director Jules Janssen, sometimes called the "father of astrophysics" and discoverer of helium. (Janssen had also studied the lines of the solar spectrum and was able to prove that the lines from oxygen resulted from oxygen in the Earth's atmosphere.) During his time in Paris, on a mid-November night in 1899 he ascended above the fog in a hot-air balloon to observe what he called "the rarest phenomenon of nature"—the Leonid meteors, of which he counted 91 that evening. And during his time in France, he also assisted with observations at another observatory on Mt. Blanc.

These personal and professional fascinations with astronomy continued when he returned to the suburbs of St. Petersburg to the Pulkovo Observatory, where he worked for many years. Debates had been raging for decades already about the "optical properties of Mars" and whether its seas were perhaps "nothing but vegetation patches" as their color appeared to change depending on the Martian seasons. Tikhov "passionately wanted to get pictures of Mars" and eagerly awaited the next opposition of 1909. It was his observations of Mars that he held were his "first step towards exploring the great problem—the problems of life on distant worlds."[6]

Citing Friedrich Engels, who had held that life was the existence of protein bodies constantly renewing their chemical constituents through nutrition and excretion, Tikhov presented himself as the well-informed materialist investigator of life, who knew that life was composed of the widely distributed elements carbon, hydrogen, oxygen, and nitrogen. He noted,

> So we, materialists, believe that life is the highest stage in the development of matter and should arise where there is a condition for this. Consequently, life is not only on Earth, but also on countless other bodies of the Universe, convenient for life.... Depending on the physical and chemical properties of the environment, these substances are clothed in a different, very diverse form.[7]

For Tikhov, the existence of life on other worlds could be inferred as a direct consequence of the principles of dialectical materialism. Given that life would emerge not only on Earth but also anywhere the conditions were appropriate, it was only fitting, he would later say, "therefore, [that] Soviet science has paid close attention to Mars, so similar to our Earth."[8]

Working with a colleague, Tikhov wanted to use the Martian opposition in 1909 and in succeeding years to confirm whether there was plant life on Mars. Although the seeds of this early astrobotanical approach were planted at the Pulkovo Observatory in 1909, it took Tikhov many years and a transplantation to Central Asia for these seeds to take root and finally flower into the Soviet science of what he would come to call "astrobotany."

In the meantime, the life of the astronomer was full of frustrations. He faced the regular difficulties of observing solar eclipses, even into the late 1920s:

> La matinée du 21 aout [1914] présente un ciel parfaitement pur. L'éclipse partielle débute. Le soleil luit dans un ciel sans nuages. L'éclipse totale est toute proche. Et soudain un cumulus surgit, s'approache du soleil et le masque. La phase complète se découvre le soleil, mais nous voyons déjà apparaître un mince croissant. Notre désappointement est complet, notre humeur exécrable! Comme pour souligner notre malchance, tous les jours suivants de notre séjour à Stavidly un ciel limpide et sans nuages luit audessus de nos têtes. Treize années entières je conservai l'amertume de cet échec, jusqu'au jour où j'obtins enfin de bons résultants à l'occasion de l'éclipse totale de 1927.
>
> Mon échec au cours de l'éclipse de 1914 me suggéra l'idée de découvrir coûte que coûte un procédé qui permetrait d'observer la couronne solaire sans attendre une éclipse.

Necessity was the mother of invention, and through these experiences and opportunities (or lack thereof), he began to envision ways to study the solar corona without needing to wait for an eclipse, while always remaining open to the next opportunity.

From 1917, Tikhov was stationed in Kiev, but with a solar eclipse soon to take place over Central Asia in 1941, the Pulkovo Observatory arranged to send an expedition to Alma-Ata (now Almaty), Kazakhstan. Due to wartime conditions, however—the war had begun on June 22—the expedition was transformed in short order into a general "evacuation of the observatory" in early August 1941. During this time, Tikhov recounts that he happened to have read in a magazine article that Tolstoy had claimed that "the ancients could hardly distinguish colors and that the Bible never speaks of the blue color of the sky." This intrigued him, and, buying a copy of the Bible, he

> quickly found several places in it that compare the clear sky with sapphire. I bought a small sapphire, investigated the spectral composition of the light passing through it, and saw that the spectrum of sapphire is very close to the spectrum of light of the clear sky, so the clear sky should be called not blue, but sapphire.[9]

This fascination with the qualities of the lights of heaven became a leitmotif of his work. He constructed a cyanometer, and it was through this work that he began to question what the color of the Earth would be as seen from the Moon. Although many science fiction authors have envisioned the Earth as seen from the Moon as some sort of "disque verdâtre," Tikhov wondered whether this would actually be the case, and he found a way to investigate this question using "ashen moonlight"—the earthshine reflected off the moon. And he then thought to apply this technique to the strange bluish or even purplish tinges of the "seas" observed on Mars that appeared to change character with the seasons and which appeared to offer no "reflection of infrared rays by its vegetative cover."[10] That Mars had vegetation of some sort was widely theorized at the time. But just what would this life be like? Tikhov wondered: "Afin d'etre en mesure de répondre à cette question qui occupa de tout temps les esprits, il importe de déterminer ce qu'est la Vie et quelles en sont les frontières."[11]

Martian Vegetation, the Invention of Analog Sites, and Soviet Politics

Confronted by this question during a presentation in Almaty on the possibility of life elsewhere in the universe, Tikhov then embarked in 1945 on a comparison of the "infrared rays of deciduous and coniferous plants." He compared pairs of plants—green oats and polar juniper, as well as birch and spruce—and discovered that the

> reflection of infrared rays from conifers—spruce and juniper—is three times less than that of birch and green oats photographed simultaneously with them. Thus, it was found that the summer green plants do not need infrared rays, so they are reflected. Given the necessity of polar juniper and spruce to absorb all available infrared rays,

Tikhov noted, these were reflected "very weakly."[12]

He believed he had at long last found a means by which to characterize the nature of Martian vegetation from here on Earth. The absence of reflected infrared rays from the seas of Mars was not a simple sign of lifelessness on Mars but, rather, might be a sign that Martian vegetation had absorbed all scarce available infrared radiation necessary for its own survival. Summarizing his long-term observations of Mars at the end of a 1945 meeting of the presidium of the Kazakh branch of the Academy of Sciences of the USSR, Tikhov later recalled: "I was lucky to be the first in the history of science to publicly pronounce the word 'astrobotany,'" which he later defined as "the science which studies vegetation on heavenly bodies."[13]

He began to develop this "new science" and built a "cohesive, friendly team of astrobotanists" who studied extreme environments on Earth—both high altitude and high latitude—as a means to more fully understanding the nature of Martian vegetation. "Life is more tenacious than we think, and can perhaps exist under completely unearthlike conditions," he said, and he repeatedly invoked a reversal of perspective between Earth and Mars—a feature common to the history of astrobiology since at least Kepler—to make his argument.[14] He also invoked possible debates among Martians as to the habitability of Earth:

> I have just imagined a fantastic scene, where Martian academicians have gathered together are debating the possibility of life on earth. A prominent Martian scientist comes forward and says, well, there could be life with such a large oxygen content in the earth's atmosphere, revealed by spectral analysis. But then again, every living creature would suffocate and burn up. . . . Does not this remind one of the discussions which are written by some of our scientist on the impossibility of life on Mars?[15]

Tikhov also foresaw that particular extreme terrestrial environments could be analogs for Martian ones: "It is quite possible that a researcher on Mars will encounter a medium no more difficult than, for example, in Antarctica," he noted. With the understanding that "Mars is one and a half times farther from the Sun than the Earth, and therefore the climate on it is harsh, resembling the climate of Yakutia and high mountains," Tikhov's astrobotanical staff "began their work with the study of the optical properties of plants in the high mountains and in arctic regions." Tikhov reported that expeditions were sent to the Zailiisky Alatau mountains close to Alma-Ata, to the Pamir mountains, "and to the cold desert of the Central Tien Shan, to the mouth of the Ob river, to Yakutia, up to the shores of the Arctic Ocean," where his astrobotanists took "spectrograms of many northern plants. In some of them—dwarf Arctic birch (*Betula nana*), reindeer moss, peduculiars, Iceland moss, and others—the spectrum, taken in July, that is, even in the warmest season of the year, did not show any noticeable main absorption band of chlorophyll."[16]

Although bands in chlorophyll demonstrating absorbed radiation were clearly visible in Tien Shan fir trees near Alma-Ata at 2°C, such bands disappeared at −6°C. Similarly, he understood "from observations made in Canada that the months of September and October are distinguished among the summer months by the abundance of light blue flowers." He concluded that these changes in spectra were a kind of external adaptive sign and that he had therefore found "a simple and natural explanation for the lack of chlorophyll absorption band in Martian vegetation" and why the "greens" on Mars have "azure, blue and even violet" tinges: "If red, orange, yellow and green light is attenuated in the reflected spectra of the plants, then azure, blue and violet light will take on greater significance." Martian vegetation would of necessity, he concluded, be bluish—just like is observed in cold climates on Earth.[17] This new science was clearly the child both of astronomy and of botany, just as Tikhov had envisioned as a young man.

In addition, as he attempted an explanation for the origins of plant colors, Tikhov referred directly the work of Ivan Vladimirovich Michurin, the peasant horticulturalist held up by the Soviet system for his practical insights into the workings of dialectical materialism in the garden and for agriculture—a practice intertwined with communist politics that came to be known as Lysenkoism. As Lysenkoism became enshrined as the politically correct state scientific ideology for biology under Stalin, the idea of the unity of the organism and its environment became firmly established in Soviet genetics.[18] Tikhov's "astrobotany" accordingly adapted itself to this new political context: "Our experiences and sayings of Michurin allowed us to draw interesting conclusions," Tikhov noted. Among other things, Soviet dialectical materialism called for explaining changes as the result of deep law-like regularities rather than as fortuitous happenstance. Tikhov referred to Michurin's experiments with heat and light in producing new varieties of roses as providing insights into not only the paleobotanical history of plant evolution on Earth but also its implications for the history of Martian flora. As Tikhov noted,

> Astrobotanists do not yet have the opportunity to visit Mars. But terrestrial plants also give the right to talk about what the vegetation on Mars is. We know that on Earth, vegetation is adapting to the environment. And this process takes place pretty quickly. By virtue of the material unity of the world, from a scientific point of view, we can assert that the law of the unity of the organism and the environment must act on Mars, just as on Earth. And, if the plants on our planet have adapted to the conditions of existence, there can be no doubt that vegetation on Mars has also adapted itself to the conditions of life. Today we have no opportunity to visit, for example, on Mars and to bring little faith incontrovertible evidence. But scientists have

> other possibilities. They can, by carefully studying the various life forms on Earth and the conditions of their existence, compare the data obtained with the conditions on the planets of the solar system and thereby make scientific assumptions about the possibility of life of organisms on other planets. This is, perhaps, the power of genuine science.[19]

Tikhov's mode of reasoning by analogy through insights gained only from observation and experimentation meant that the blue seas of Mars were biosignatures of Martian vegetation adapted to exist in a cold, dry Martian environment. He clearly and forthrightly notes the analytical moves that led him to this conclusion:

> Investigating the possibility of a plant world on Mars, we had to, as it were, descend from Mars back to earth to study the optical features of terrestrial vegetation, in order to then once more refer back to Mars and say what kind of earthly plants are most closely resembled by the plant cover of some particular section of the "seas" of Mars.[20]

This kind of transportation from the heavenly to terrestrial and back again—what Tikhov called "the cosmic role of plants"—was of course an extension of Timiryazev's own earlier understanding of photosynthesis, as Tikhov noted. But it is also a move quite similar to moves of analysis about terrestrial analogs made in astrobiology today.

What is distinctly Soviet about Tikhov's effort was the way in which he repeatedly insisted on the practical agricultural implications of his astrobotanical research for developing frost-resistance and drought-hardiness in plants on Earth. This unity and interpenetration of opposites—this marriage of dialectical materialism with spectroscopy, and of Earth with Mars—he held, "poses many interesting problems for biology and biophysics, which will help to unravel the secrets of life in general." In so doing, Tikhov took Soviet Michurinism and placed it in an interplanetary context. This also enabled Tikhov to appeal to dialectical materialism as the solid philosophical and politically correct base from which to challenge what he called *biological geocentrism*: "the statement that the Earth is an exemplary, most favorable body for life." Just because "the physical conditions on other planets are significantly different from those on earth," Tikhov argued, is not grounds for dismissing the universality of life.[21] Tikhov himself held that

> in the future, as Soviet science and technology will present the most powerful instruments for observation, [and] truly unlimited horizons will open before the astrobiologists. Mankind will disclose the secrets of life on other planets. The study of life on earth and on other planets are tightly interconnected.

Emerging and flourishing in the harsh political and natural climate of Soviet Central Asia, astrobotany soon was expanded to include the study "of microorganic life on the giant planets" and by 1952 was called "astrobiology."[22]

Tikhov's vision was inspiring to many Soviet investigators. By the end of February 1953, one major conference adopted the resolution that

> only the Soviet Union hoisted the flag of progressive cosmic biology high and widely developed fruitful work in this complex scientific field. Astrobiology opens up ways of creative collaboration between the members of many sciences, astronomy, physics, mathematics, biogeochemistry, paleobotany, geology, geography and others.[23]

Conclusion

Finding new ways in which earthly politics and culture are both constitutive of and exported into our astrobiological understandings of extraterrestrial elsewheres is an essential part of any proper [humanistic approach to the study of astrobiology, and the case of Tikhov's astrobotany highlights several important features. We have already seen several key features of Tikhov's analytical moves: from studying earthshine to studying plantshine as a kind of biosignature able to be analyzed at a distance, and proposing the study of extreme environments—high latitude and high altitude—to better question whether temperate zone life forms are to be privileged in our analysis (an early argument for extremophiles, perhaps). He also routinely reversed perspective between Earth and Mars to see what we might learn about the Earth from how it might be seen from elsewhere, and to learn about Mars from what we might learn here on Earth. (He also used this reversal of perspective to disarm critics.) Thinking back and forth off the planet and reversing perspective is far from a unique feature of Soviet astrobotany—indeed, it is a common feature of astrobiology, from Kepler's *Somnium* to Sagan's proposal to turn cameras back toward Earth to try to detect a biosignature—and is even a contemporary feature of astrobiology today, as we struggle to understand how best to analyze an exoplanet's atmosphere for signs of life.

But the reversal of perspective was not only an unproblematic analytical tool of science, declaring the supremacy of dialectical materialism for understanding life in the universe—but also could serve as a tool of cultural critique. Tikhov's invocation of a Martian perspective on an oxygen-rich Earth and its likely uninhabibility had its own analog in the capitalist West, in a contemporaneous work from the 1950s that was written from the American perspective—or rather, it claimed, from the Martian perspective of Earth. *Is There Life on Earth?* purports to be a report to the Congress of Mars, but it is illustrated by a cartoonist for *The New Yorker* and contains delightful humorous takes on modern terrestrial life as observed by Martians.[24] Chief among these was the recognition of Martians studying earthly forms of life that there are two countries on planet Earth easily identifiable by the signs all over each: One country is called "Coca-Cola" and the other "vodka." Although every historian worth their salt knows that our views of life in space are also always in some way views about us here on Earth, a perspective that Tikhov literalized in his work—"descend[ing] from Mars back to earth to study the optical features of terrestrial vegetation, in order to then once more refer back to Mars"[25]—this book takes that alien reverse perspective to new heights.

What is also clear is that Tikhov tied his criticism of looking for the wrong kind of life elsewhere—his argument for studying extremophile life, we might say—with the basic principles of dialectical materialism and a critique of "biological geocentricism in contemporary biology." He wrote,

> Hence, we should not imagine the life conditions on other planets to be under the same physical conditions as those on earth, and absolutely not those in the large cities where scientists mostly take their studies of animal, and particularly, of plant physiology. All these investigations were recorded with the thought that the life conditions in these localities were most favorable to life. It appears to me that it is about time to cast off this biological topocentricism, or, in a more general sense, biological geocentricism.[26]

In this most revealing passage, Tikhov's study of what he called the "localities most favorable to life" must be understood not only in terms of biology but also in terms of the conditions of existence for such biological study: urban or rural? Bourgeois or peasant? Although much extant scholarship on Tikhov—and there is not much—is quick to highlight his coining of "astrobiology" in mid-century, amid a variety of other contenders in other countries and languages, or the many ways in which his work presages later or even contemporary questions in astrobiology today, such as how studies of "plantshine" have continued apace,[27] the political valences and contexts of his work have often been ignored. But what is clear is that Tikhov's texts speaks in all these registers simultaneously—he moves seamlessly from an analysis of the conditions for life in various places on Earth, on other celestial bodies, and the social ground and contexts for producing this work in the first place. In so doing, Tikhov was surely dealing with the political and social realities of his moment. But we might also consider whether our analysis of his contributions suggests that we properly complicate our understandings not only of *where* life might be found and *how* it might adapt but also of where and how life is studied, and how these frameworks and limits construct our understanding of life. For Tikhov, working during a time of Lysenkoism, the perspective of the Soviet peasant farmer, Michurin, was invoked as a privileged perspective for understanding what was most favorable to life. What privileged perspectives are we invoking today? What might prove instructive from a recollection of this generally overlooked and forgotten Soviet lineage in the history of astrobiology?

Tikhov's adherence, whether real or staged, to Michurinist reinterpretations of principles of dialectical of materialism was difficult for Americans to interpret—even Americans who had paid good money to have some of Tikhov's works translated for their own professional development, as Carl Sagan had done. Like Tikhov, for whom dialectical materialism meant that life would emerge wherever conditions would permit, Sagan held in 1959,

> We have now an increasingly plausible picture of the steps where by life evolved on earth. This supports a strong expectation of parallel developments elsewhere, whenever the availability of carbon compounds (which are universal), water or other solvents, and a suitable temperature range are compatible with the evolution of chemical complexity.[28]

And, a few years after Tikhov's death in 1960, Sagan reported in 1967 that some Soviets claimed

> that dialectical materialism requires life to be present on all planets. They were looking forward to finding it everywhere. However, this had the annoying implication that if life was found to be absent from any planet, dialectical materialism was thereby disproved.

(This is, of course, false, but Sagan felt that debates over the exobiological implications of dialectical materialism in the Soviet Union occurred in the United States "under somewhat more openly religious guises.")[29] This overlap between plants, planets, and the political is apparent even in the fact that among Sagan's collected papers and translation of Tikhov's work is a copy of the pseudonymously published "Is There Life on Mars?" (*Есть ли жизнь на Марсе?*), which despite sounding almost like Tikhov's "Is Life Possible on Other Planets?" is actually a fictionalized account of anti-Semitic persecution involving the "Interplanetary Lunar conspiracy, Chaim, Abram & Co."[30] Even in going to Mars, one never leaves Earthly troubles behind.

But Tikhov's critique of biological geocentricism was echoed and amplified by prominent exobiologists engaged in their own efforts to expand the category of life. As Joshua Lederberg asked in 1959, "I hope we don't fall into the geocentric trap of assuming that the planetary biota will be classifiable in a terrestrial taxonomy—are the Martian 'lichens' supposed to be 'symbiotic associations of algae and fungi'?"[31] Arguing for earthly analogs to understand extraterrestrial elsewheres ran the risk of unreasonably equating organisms from different planetary trees of life.

A couple of generations later in the history of astrobiology, politics would intervene in a new way: As Jim Strick and Steve Dick have shown, American exobiology would be rechristened "astrobiology" by NASA in response to congressional criticisms of the waste of taxpayers' money searching for "little green men." And so while Tikhov viewed astrobiology as an eminently *practical* science, tying together plants and planets as a political necessity of his time and place, NASA Astrobiology's Director, Barry Blumberg, created and funded the astrobiology chair at the Library of Congress precisely because he thought that astrobiology was the perfect example of a pure science *without* immediate application—and that in so doing, it could serve as a model for members of Congress seeking to understand the importance of basic and not always applied research. The explicitly political contexts of astrobiology, and their remarkable inversions over time or national boundaries—Soviet, American, French—are, or should be, I suggest, inescapably central features of any effort to understand the history of astrobiology.

3

Styles of Contemporary Space Exploration

Columbian and Vespuccian Modes of Researching Alien Worlds

Istvan Praet

Comparing Outer Spaces: Introduction

The pluralization of outer space is a characteristic yet barely noticed feature of the present day and age, marking it off from an earlier Space Age, the half-century period that runs, roughly, from 1960 to 2010. In Chapter 1, the anthropologists Lisa Messeri and Valerie Olson consistently and self-consciously speak about *outer spaces*. The significance of that move cannot be overemphasized. It is the hallmark of outer space studies in the sense outlined throughout this volume. This pluralization is not just a conceptual advance at the cutting edge of social scientific research, however. Messeri and Olson primarily present it as an injunction, as a path they urge their colleagues to explore further. In this chapter, I suggest it can/must also be grasped as a description—pluralization describes something that is unfolding right now in the planetary sciences and that has by and large remained under the radar. A new, previously unimaginable horizon opens: the comparative study of outer spaces. In outer space studies, there is an emerging consensus that the modern notion of "outer space" we ended up with in the early Space Age is an unexpectedly provincial construct, its vast reach notwithstanding. Nowadays, it is customary to accept that the extraterrestrial begins 62 miles above mean sea level—at the von Kármán line, as the edge of space is technically known.

Yet other ways of delimiting what constitutes "outer" are imaginable too. In fact, other varieties of outer have been imagined on a number of occasions throughout history. That is why the input of historians of science is so crucial for outer space studies. Specifically, I think of the work of

Istvan Praet, *Styles of Contemporary Space Exploration*. In: *Otherwhere Ethnography*. Edited by: Istvan Praet and Perig Pitrou, Oxford University Press. © Oxford University Press (2025).
DOI: 10.1093/9780197790885.003.0004

John Tresch (2012), about utopian science in 19th-century France, and that of Jessica Riskin (2016), on automata and the banning of agency in the modern life sciences. What their respective monographs, *The Romantic Machine* and *The Restless Clock*, have in common—despite the very different subjects treated—is that they develop the idea of "roads not taken," of "foreclosed ways of thinking" which resonate with current scientific trends and may, seemingly against all odds, even make a comeback. Just as modern certainties about the separation between art and science (Tresch) and passive-mechanism (Riskin) may pass, the modern Space Age conception of outer space is not set in stone. In the space sciences, other, long overruled ways of conceiving "outer" may be about to return too—that is what this chapter aims to show.

The intention here is not so much to debunk modern scientific cosmology as to portray its fundamental partiality: As the anthropologists Sophie Houdart and Christine Jungen (2015) have pointed out, there are multiple ways of connecting with the cosmos. And as speculative philosophy teaches us, the modern belief in a single connection, in one supremely objective road toward "the cosmos as it really is," is a bit of a mirage (e.g., Stengers 2011; Debaise 2017). Outer space remains and has always been an intrinsically abstract concept—to believe that the most advanced sciences have direct access to it is falling in the trap that the founder of speculative philosophy, Alfred N. Whitehead (2011 [1925]), famously called "the fallacy of misplaced concreteness." Its ongoing exploration is typically presented as a feat of rocket engineering rather than as an enterprise of conceptual creativity. In what follows, I take a different route and approach space exploration not so much as a technological challenge as an adventure of ideas, to recycle one of Whitehead's (1967 [1933]) favorite phrases. Following the lead of speculative philosophers, I argue that many of those who do research into alien worlds are not just talented scientists but innovative metaphysicians too. Space agencies such as the National Aeronautics and Space Administration (NASA) and European Space Agency are not just scientific nerve centers: They are philosophical hotspots as well, places where modern habits of thought are being reformatted, even though that specific role has so far barely been acknowledged. My suggestion is that metaphysical style is at the heart of contemporary space exploration.

In Chapter 1, Messeri and Olson introduced outer space studies as an eminently cross-disciplinary endeavor. In the early 2000s, their colleague Debbora Battaglia (2006, 2) had already noted that one of the most striking features of the idea of the extraterrestrial is "the extent to which conventionally distinct fields of knowledge cross-connect, collide, or pass through one another under its influence." The anthropologist Stefan Helmreich (2016, 91–92), for his part, likes to think of it as an "adisciplinary" or indeed "undisciplinary" endeavor. The present chapter, which is informed by ethnographic research with astrobiologists and planetary scientists, illustrates that observation. Although my specific recipe of cross-disciplinarity, which mixes social anthropology, history of science, and speculative philosophy, is to some extent idiosyncratic, the general idea on which the chapter is premised is very much in alignment with what others in this volume propose. Concretely, I compare three outer spaces: the ocean sea, the interplanetary medium, and outer time.

My choice to foreground this at first glimpse rather strange threefold requires a little explanation. In modern cosmology, the interplanetary medium, as the space between planets in the solar system is technically known, fulfills a special role; it is not just any boundary. One can say it is apprehended as the *boundary of boundaries*. In the sense that the tenuous space between planets is the only boundary deemed capable of separating the familiar from the alien, it is the only truly *serious* boundary that modern people recognize. If you want to find extraterrestrial life, crossing the interplanetary medium is a minimum requirement. Encountering aliens requires voyaging from our home planet to a not-so-homey elsewhere. In the modern template, outer space is a uniquely serious boundary, whereas all other boundaries are distinctly unserious insofar as they mark partial rather than total discontinuities.

Take the example of oceans: When one crosses the Atlantic, one will reach another continent but not an alien world, so moderns believe. Latter-day Europeans deem it self-evident that Americans are not aliens. Americans may be different but—so it is presumed—only just a little bit. Likewise, latter-day Europeans rarely bother to query whether macaws, peccaries, or jaguars are aliens, if at all. Ultimately, the American fauna are cast as familiar, as related to their Old World counterparts. For those who operate under the rules of modern metaphysics, it has become too obvious to even mention: Since the advent of evolutionary theory, humanity and terrestrial life as a whole are conceived of as one big family. Unlike spacefarers, contemporary seafarers are constricted to encountering more of the same: other humans, other animals, other plants. As a boundary, the ocean is *unserious*: It separates two entities that are taken to be constitutionally similar. What tends to be underestimated, however, is that this particular way of conceptualizing the ocean only stabilized relatively recently. One only needs to go back to the "ocean sea" (*mare oceanum*) of Renaissance days to become aware that the current setup was not inevitable. As we will see, Christopher Columbus, whom the Catholic Monarchs of Spain had proclaimed "admiral of the ocean sea," already viewed the Atlantic as unserious, but in the so-called epoch of the great discoveries, the metaphysical status of the ocean initially remained unsettled. If Amerigo Vespucci—the navigator after whom America was named—would have had his way, today's default metaphysical setup might have looked rather differently: We will see that Vespucci, unlike Columbus, imagined the ocean as a *serious* boundary, along the same lines as moderns conceive of the interplanetary medium.

Or consider geological distances, which always mark a partial discontinuity in the modern setup, never a total discontinuity. After all, the Earth is conventionally conceived of as one continuous planet with an approximate age of 4.5 billion years. Modern geologists, one can say, have a particular fixation with continuity. They do recognize a couple of major ruptures, to be sure, but these are never envisaged as serious boundaries. In their mindset, the Great Oxidation Event was momentous, but at the same time they assume it did not alter the basic nature of our planet. Possessing a thoroughly oxygenated atmosphere is merely a secondary quality, it is supposed: The previous, oxygen-less Earth was still an Earth, not a separate, alien planet. Modern biologists have an equally strong fixation with continuity. The meteorite impact that caused the demise of the dinosaurs was literally earthshaking, yet it is never conceived of as a serious boundary. Evolutionary biologists typically agree that the Cretaceous world of the dinosaurs was a different Earth but not an alien planet. In the modern sciences, planets are only allowed to evolve; planetary metamorphosis has been ruled out by decree. Along the terrestrial timeline, there may be partial discontinuities, but total discontinuity has de facto been forbidden. Global catastrophes and mass extinctions are never attributed the same capacity of total separation as outer space. Under the rules of modern metaphysics, no temporal boundary marks a clean break. That privilege is restricted to the one spatial boundary that has been proclaimed the boundary supreme—the interplanetary medium. Moderns believe in outer space but—curiously and rather incoherently—not in outer time.

Yet we may be witnessing a metaphysical sea change at present. An increasing number of scientists reimagine geological and evolutionary distances as serious boundaries. That is why, in a field such as astrobiology, the ancient

Earth is sometimes pictured as an alien planet or indeed as a sequence of multiple, alien planets. Conversely, we will see that outer space itself is sometimes rethought as an unserious boundary nowadays. Celestial bodies that potentially harbor life, such as Mars or Jupiter's moon Europa, are frequently presented as "the Earth's cousins"—that is, as familiar rather than as alien planets. Basically a comparison between three outer spaces, the chapter suggests that the old notion of the ocean sea, the Space Age notion of the interplanetary medium, and the newly emerging notion of outer time have a great deal in common.

The Ocean Sea

The modern imagination was durably shaped by a narrative that foregrounds the triumph of adventurous sailors over bookish scholars. The story is neat and simple, and it runs as follows. In times past, the inhabited world was believed to consist of only three major landmasses, centered around Jerusalem: Europe, Africa, and Asia. This is the terrestrial ecumene depicted in medieval *mappae mundi*. Then, in the late 15th century, that ancient worldview was shaken. As the likes of Christopher Columbus (1451–1506) and Amerigo Vespucci (1454–1512) explored the high seas, more accurate world maps began to be drawn, free from century-old prejudices. In that conventional story, Columbus and Vespucci are the harbingers of the empirical method, which would come to characterize modern science. By means of their firsthand observations, they challenged the sacrosanct authority of classical books such as Ptolemy's *Geographia*. Their voyages announced the ascendency of a more realistic outlook over one that was steeped in persistent errors. In no small measure, sailors guided us from premodern obscurity to modern enlightenment, from subjectivity to objectivity. This well-rehearsed narrative may be convincing but is in fact deeply flawed, as anyone who is up to date with the history of Western knowledge will attest. Far from impartial observers, figures such as Columbus and Vespucci were products of their Renaissance milieu. Besides glass beads and hawk's bells, or parrots and slaves, their caravels also carried metaphysical cargo. The letters they wrote to their royal patrons and the log books they kept were neither radically distinct from extant cosmographies nor uninfluenced by the scholarly context of their day. Theirs was an epoch where science itself was understood as "wordy science" rather than as "natural science," and where the divine revelations of the holy scriptures and the wisdom enclosed in classical texts carried greater weight than direct observations by fallible humans. In the immediate aftermath of the so-called great discoveries, the Bible still took precedence over the truths one might have hoped to uncover in nature. Whatever precisely happened, it was neither the gradual rise of empiricism nor the inexorable advent of modernity.

Upon closer inspection, it turns out that Columbus and Vespucci had quite distinctive ways of apprehending the distant shores they explored and rather opposing ways of picturing their inhabitants. They also had contrasting manners of conceptualizing the "ocean sea" they so skillfully navigated. One can say that each of them had his own, unmistakable metaphysical style. The historian of humanistic scholarship, Anthony Grafton (1992, 84–85), has identified a key feature of these two styles: Whereas Columbus was inclined to emphasize similarities and continuities, Vespucci was disposed toward differences and discontinuities. In brief, Columbus familiarizes and Vespucci alienates. Grafton has demonstrated that neither of them sought to assert the superiority of firsthand observation to ancient, authoritative texts. Both were also "men of the book," albeit men with varying literary preferences. Columbus had a penchant for authors such as Aristotle and Saint Augustine, and he shared their belief that the surface of the world mostly consisted of land rather than water. He was particularly influenced by the astrological and eschatological writings of a French cardinal, Pierre d'Ailly (1351–1420) (Moffitt Watts 1985). Columbus' millennial worldview is borne out in his "book of prophecies," a collection of quotations from the Bible and from ancient, patristic, and medieval sources, which he compiled between his third and fourth voyages. Vespucci was influenced by classical authors such as Herodotos and Pliny the Elder, who rendered the antipodes—those who lived on the other side of the globe—as barbarous and utterly alien. In line with his humanist education, he emulated these authors in his letters: He did not refute them or attempt to surpass them, as is often claimed, but sought to "update" them through rhetorical eloquence and literary ingenuity, following the guidelines of Petrarch (Van Haeringen 2020). As the first chief-pilot (*piloto mayor*) of the *Casa de la Contratación* in Seville, he promoted a style of reasoning that combined cosmographical theory with navigational experience. The institution was a "veritable chamber of knowledge," the Jet Propulsion Laboratory of its day: Among other things, the chief-pilot was responsible for the improvement of nautical and astronomical instruments and for the continuous updating of the royal portolan, the master sea chart from which all particular charts had to be drawn (Barrera-Osorio 2006). What is of import here is the remarkable contrast between the thinking of the admiral of the ocean sea and that of the chief-pilot. Grafton (1992, 85) notes, "Vespucci's insistence on absolute difference was as literary as Columbus' insistence on familiarity." Crucially, I add, their respective methods of boundary-conception were radically dissimilar. If Vespucci took the ocean to be a serious boundary, Columbus conceived it as fundamentally unserious. Of course, that is not to say that Columbus took the dangers of navigating the high seas lightly or that he underestimated the hardships and the manifold obstacles that came with long-distance marine voyages. That he viewed the world and the extent of its oceans as smaller than conventionally assumed at the time is not the point. And that Vespucci viewed the world's watery surface as larger than most of his contemporaries is not the point either. The serious/unserious distinction is used in a technical, metaphysical sense here, and has nothing to do with absolute size or distance.

A basic, preliminary definition suffices for my purposes here: Crossing a serious boundary brings one into an unfamiliar, alien world, whereas stepping (or sailing) across an unserious boundary just leads to more of the same. To suggest that Columbus was committed to "more of the same" may seem off the mark at first blush. Yet one must remember that Columbus' claim to fame—by his own account—was not to have discovered a new continent but, rather, to have *ascended a nipple*. To the end of his days, he was convinced that he opened a new, western route to the Indies and to Cathay—that is, to one of the three known continents. America, as a separate continent in its own right, was not imaginable for him (Paine 1995). Unlike Vespucci, he never made any claim to novelty in that regard. He always stayed within the canon of classical and biblical cosmography. As the anthropologist Carol Delaney (2006, 287) has shown, Columbus did not so much envisage his voyages as "discoveries" than as "revelations"—they portended an impending, final crusade. Columbus was steeped in Franciscan Catholicism, and his mindset was apocalyptic through and through. The end of the world was nigh. Everything he did had been foretold. It was a first step in the accomplishment of a divine prophecy: The evangelization of the Indies and the gold the Spaniards hoped to acquire (which he associated with Salomon's mines and the biblical Ophir) were meant to serve the Christian reconquest of Jerusalem and, ultimately, the second coming of Christ. At any rate, Columbus proudly claimed to be the first European to locate the precincts of a giant, planetary nipple. When he observed the enormous amounts of sweet water that flowed from the mouth of the Orinoco River into the sea, he interpreted this as a sign of the proximity of the terrestrial paradise rather than as an indication for the existence of an unknown southern landmass. On the basis of stellar observations conducted during his journeys, he had already come to the conclusion that Earth was not entirely spherical but pear-shaped. One hemisphere was round as Ptolemy described it, but the other hemisphere, which the ancients did not know, resembled "a woman's nipple on a round ball" (Cohen 1969, 218). The terrestrial paradise, he conjectured, was most likely situated on that protruding nipple or pear stalk, which itself was located at the eastern end of the Asiatic mainland.

However ingenious and original this line of reasoning was, it disturbed neither the classical world picture of three principal landmasses surrounded by a smallish ocean nor the traditional Christian template of a terrestrial paradise at the eastern extremity of the world. From a Columbian perspective, the newfound "islands" may have looked a little strange, but they were never totally alien—hence typical turns of phrase such as "the land and the trees were very green and as lovely as the orchards of Valencia in April" (Cohen 1969, 219). That basic familiarity also applies to the native inhabitants, whom he often portrays as quick-witted, industrious, and comely: "I have not found the human monsters which many people expected. On the contrary, the whole population is very well made" (121). When some of his crew members wanted to throw the Indian passengers overboard when food shortage became an issue during his second return voyage, Columbus prevented this because—in the words of his son and biographer Ferdinand—"he considered them as kindred and fellow Christians and held that they should be no worse treated than anyone else" (199). The genealogical view according to which the whole world was settled by the descendants of Noah was widely held at the time, and Ferdinand's testimony is one among several indications to believe that Columbus adhered to some form of it, even though the admiral of the ocean sea never explicitly mentions the biblical patriarch in any of his letters. In any case, what seems certain is that he did not consider the Indians to be beyond the pale: They could be brought into the Christian fold with minimal effort.

Now compare this to Vespucci, who in a letter to his Florentine patron described the natives in the following terms: "They were of a race called Cannibals, for almost the majority of this race, if not all, live off human flesh: and of this fact Your Magnificence can be certain. . . . Many times we saw the bones and heads of some of those they had eaten"

(Formisano 1992, 9). The chief-pilot was imbued with humanistic scholarship, as testified by the Lucretian overtones of his writings (Brown 2010, 31). In contrast to Columbus, Vespucci delighted in describing exotic particulars and macabre details. The naked natives pierce their lips and cheeks with bones and colored stones "as big as Catalan plums" (32). In their houses, he reported, they hang salted human carcasses from the rafters, "much as we hang up bacon and pork." They are cruel, do not obey any king, and do not know private property (33). Even more shockingly, they eat on the ground "without tablecloth or any other sort of napkin" (63). The birds and the wild animals are exceedingly foreign too, according to Vespucci. There are so many kinds of "dreadful and ill-formed beasts" that he doubts whether they all would have fitted into Noah's ark (31). It is by reasoning along the same, alienating lines—emulating the "hard primitivism" of Lucretius—that he came to a much more consequential conclusion, namely that he and Columbus had landed on an independent continent that is not Asia. Vespucci fathomed that they had found a totally unfamiliar landmass—a *Mundus Novus.* Columbus, as I noted previously, always maintained he merely reconnoitered new routes in an already familiar world. Unlike the latter, who never conceived of the newness of the regions his fleet had reached, Vespucci grasped that what would soon become known as "America" formed a truly alien entity unknown to the ancients—a veritable new world. And he was the first to realize that Europe and Asia were not separated by one but by two oceans.

The point of telling all this is not to weigh in on futile debates about precedence. With the benefit of hindsight, it is typically presumed that Vespucci was "right" and Columbus "wrong." Yet off-the-shelf conclusions like this do not bring us very far, for they merely repeat the platitudinous template of the unavoidable triumph of empiricism. Instead of praising Vespucci as the herald of modern geography and as the precursor of the natural scientific outlook more generally, it is more apt to reflect on Grafton's (1992, 83) observation that "even more than Columbus, Vespucci proved to be a man of the book." In other words, one needs to appreciate that the late 15th-century discovery of America the continent, at least its conceptual discovery, was very much an achievement of bookish, wordy science. The appearance of "America" hinged not only on acute observations of the physical world but also on a particular metaphysical mindset. It depended on navigators' effective use of quadrants and astrolabes, to be sure, but that was not a sufficient condition. "America" arguably could not have emerged without the adoption of a metaphysics oriented toward difference—of a Vespuccian-style metaphysics that made the alien thinkable. In an important sense, America the continent was also a creative decision and a product of specific literary habits. What is more, this insight does not just pertain to peregrinations of the past. It has profound implications for our understanding of present endeavors of exploration too. To quote Grafton one more time, "We would be deluded to think that modern travellers are more alert—or less prone to impose on what they find the values and visions they bring with them—than the self-consciously alert Vespucci" (85).

The Interplanetary Medium

Why moderns got so infatuated with the interplanetary medium and elevated it to their boundary of boundaries is an unresolved puzzle that merits much greater attention from social scientists and humanities scholars than it has received so far. After all, the space between the planets is rather insubstantial as far as boundaries go—an average garden fence or an ordinary brick wall are arguably more secure. Granted, it is not a veritable vacuum. It consists of dust particles in low densities, includes cosmic rays, and behaves as a plasma. It carries the Sun's magnetic field with it and is electrically conductive. Yet however tenuous this near-vacuum may be, modern cosmology attributes it the most weighty role thinkable: separating "here" from "elsewhere," dividing the terrestrial from the extraterrestrial. The interplanetary medium is envisaged as a total discontinuity: Fare across and you reach alien worlds. Consider the examples of Mars and of two Jovian moons, Io and Europa. If the Earth is an aqua planet, Mars is usually envisaged as a cold desert planet and as an alien dune world: Its surface is partly covered with exotic sand dunes whose shape and motion—as the planetary scientist Matthew Chojnacki has shown—are not so much determined by aeolian processes as by factors that do not influence terrestrial sand movement (Chojnacki et al. 2011). Io is a small but dense moon whose surface is dotted with hundreds of highly active volcanoes that do not take the form of protruding cones but of cauldron-like depressions known as *paterae*. These "alien volcanoes," as NASA volcanologist Rosaly Lopes refers to them, are pits rather than mountains (Lopes and Carroll 2008). They confound what we think of as geysers and volcanoes here on Earth: They spout umbrella-shaped plumes of sulfur-rich gases and dust that continuously "spray-paint" the moon's surface in a wild palette of bright colors. Europa is the favorite moon of many astrobiologists because it is believed to possess a global ocean beneath its icy outer shell. And where there is liquid water, there might be life. Yet Europan life forms would be very different from life-as-we-know-it, researchers emphasize. Because the subsurface ocean is in all likelihood devoid of direct sunlight, photosynthesis appears impossible. Europan life may be powered in other ways though. Undersea volcanoes may play the role of miniature suns, supporting chemosynthetic (rather than photosynthetic) ecosystems. Some astrobiologists even speculate that Europan life forms will have developed bioluminescence. What we can be sure of, planetary scientist Kevin Hand (2020) emphasizes, is that Europa's ocean truly is an *alien ocean* and that the life it possibly contains will be truly alien as well. In summary, Mars, Io, and Europa epitomize alienness. Time and again, the interplanetary medium is implicitly conceived of as a total discontinuity.

The present-day exploration of outer space echoes the past exploration of America by European sailors in a number of ways. The outward appearance of the characters may have changed, but the plot itself remains curiously unchanged. Nowadays, the protagonists are alien dunes, alien volcanoes, and alien oceans instead of cannibals, natives who have piercings as big as Catalan plums, and brutes who ignore the usage of tablecloths. When one goes through the writings of the aforementioned scientists, one cannot help but hear the voice of Amerigo Vespucci all the time. *These alien dunes do not use napkins. These alien volcanoes are monstrous. Unlike ours, this alien ocean is not Christian.* Today, the space sciences are premised on a metaphysics oriented toward difference, just as the geography practiced by the Florentine sailor who gave his name to America. In the exploration of outer space, the default setting is Vespuccian, not Columbian, with rare exceptions. One exception relates to the panspermia hypothesis: If Martian life was seeded from Earth (or vice versa), Martians would be our biological cousins rather than veritable aliens. This is a characteristically Columbian line of thought, which echoes the Genovese admiral's inclination to view the native inhabitants of America as the descendants of one of the lost tribes of Israel rather than as an unrelated form of humanity. By and large, however, the interplanetary medium is conceived of along the same lines as the Vespuccian ocean sea: as a radical boundary that separates the familiar from the alien. Today we may be witnessing an about-turn, however: the emergence of a Columbian style of space exploration. Consider the following observations.

Martian dunes, some planetary scientists advance, may be more familiar than hitherto assumed. It is not just that they resemble terrestrial dunes in some ways: They even resemble terrestrial organisms. The 20th-century dune expert Ralph A. Bagnold already noted that terrestrial dunes behave very much like living things: He mused about their capacity to keep their precise shape and to repair any damage as they migrate downwind (Bagnold 1941). As they move, they absorb "nourishment" and they "grow." They even create "babies" just like themselves, which then run ahead of their "parents." Extraterrestrial dune experts have observed similar behaviors on Mars. Barchans (i.e., crescent-shaped dunes) can coalesce, cleave, and sometimes "breed" in the sense that little dunes are formed at the horns (Lorenz and Zimbelman 2014). At a 2013 planetary science conference in London, one of the participants told me, "If you could speed up aerial footage these isolated barchans, their arms outstretched in the direction of movement, would appear to migrate over the surface as a bizarre herd of animals—a kind of Martian Serengeti!" Rather than alienating extraterrestrial barchan dunes, he familiarized them as gazelles, zebras, and wildebeest. In how far my London interlocutor is representative of his wider profession I cannot tell, but he certainly is not unique. When NASA's *Mars Reconnaissance Orbiter* recently released an image of a large crater that functions as a sand trap, scientists floated a similar idea. On the image of the crater one sees a group of barchan dunes moving ahead in a V formation, "just like migrating birds." Here, extraterrestrial dunes are familiarized as geese. To envisage dunes as living things goes against the (Vespuccian) grain: It exemplifies a Columbian style of space exploration. Mars itself becomes less dead, more interesting, less alien, more Earth-like. *These Martian dunes are Christians—don't throw them overboard.*

Similarly, Io's volcanoes may not be as alien as they appear at first sight. Some planetary scientists believe that the moon's super-hot lava flows offer a window on the volcanic processes that characterized our own planet more than 2 billion years ago. NASA scientist Torrence Johnson imagines the Jovian moon as a time capsule of sorts: "Io is the next best thing to traveling back in time to Earth's earlier years. It gives us an opportunity to watch, in action, phenomena long dead in the rest of the solar system."[1] The planetary geologist Tracy Gregg concurs: "When we look at Io, we may be seeing what Earth looked like when it was in its earliest stages, akin to what a newborn baby looks like in the first few seconds following birth."[2] The image Gregg uses is striking: Io is no longer pictured as the epitome of celestial alienness, but as the Earth's younger cousin or indeed its little baby brother. Upon closer inspection, many of its more exotic features turn out to be not that exotic after all. Apart from the fact the Ionian sulfur plays the role of terrestrial water, its geochemistry is actually surprisingly similar to that of the Earth (Battaglia et al. 2014). Giant lava lakes such as Loki Patera are not unlike anything else, Gregg points out: They could be "an Ionian version of [terrestrial] mid-ocean ridges."[3] It is true that pateras such as Loki are approximately circular and that mid-ocean ridges on Earth are long and narrow, but these geological features are essentially alike: Their different form is merely a consequence of the respective absence and presence of plate tectonics.[4] One should also keep in mind that the Earth itself only developed plate tectonics after a couple of hundred million years. As Io is rethought as familiar rather than alien vis-à-vis the Earth, the interplanetary medium is redefined as an unserious boundary akin to Columbus' ocean sea. One can say that scientists like Johnson and Gregg are devising a Columbian style of space exploration. *I have not found the volcanic monsters which many people expected.*

Finally, Europa may not be that alien too. At the 2020 Goldschmidt conference, NASA geochemists announced that its subsurface ocean is in all likelihood much more similar to terrestrial open-air oceans than previously realized.[5] According to the models of planetary scientist Mohit Melwani Daswani, Europa's ocean would have been moderately acidic at first, with significant concentrations of carbon dioxide, calcium, and sulfate. Through metamorphic processes, it then gradually became chloride-rich, a conclusion that is corroborated by findings from the Hubble Space Telescope—traces of chloride have been detected on the moon's frozen surface. If this line of reasoning is correct, the composition of the encapsulated lunar ocean resembles oceans on Earth to a remarkable degree. And this has implications for astrobiology. "We believe," Melwani Daswani notes, "that this ocean could be quite habitable for life."[6] In summary, an alien ocean turns out to be unexpectedly familiar. In 2016, NASA already announced that Jupiter's icy moon is "surprisingly more Earth-like than previously thought" because it is believed to have not only an ocean of salty water but also a rocky mantle, and an iron core.[7] Even its most obviously un-Earth-like feature, the global shell of ice that envelops it, may be more familiar than conventionally thought. Planetary scientists Simon Kattenhorn and Louise Prockter (2014) have established that the brittle outer layer constitutes a global patchwork of tectonic plates, complete with ridges and subduction zones, thus defying the long-standing idea that the Earth is the only body in the solar system with active plate tectonics (Selvans 2014). Scientists such as Melwani Daswani, Kattenhorn, and Prockter are the trailblazers of a Columbian style of space exploration. The interplanetary medium is de facto transformed into an unserious boundary in their hands, just like Columbus' ocean sea. *This planet and its ocean are civilized and industrious, just like ours.*

Outer Time

In contrast to the space sciences, modern geology has adopted a Columbian mode of exploration. Along the terrestrial timeline there may be partial discontinuities, but total discontinuity has de facto been forbidden. Just like Columbus' ocean sea, global catastrophes and major extinction events are a priori defined as unserious boundaries. Geological and evolutionary expansions of time, even if they stretch over millions of years, are never attributed the same capacity of total separation as outer space. Under the rules of modern metaphysics, no temporal boundary marks a clean break. That privilege is restricted to the one spatial boundary that has been proclaimed the boundary supreme—the interplanetary medium. As I stated at the outset, moderns believe in outer space but—curiously and rather incoherently—not in outer time. In this section, I describe the emergence of a metaphysical countercurrent. Specifically, I focus on the rise of discontinuity thinking within astrobiology and on the concomitant birth of a concept of outer time. What we are witnessing, I suggest, is the appearance of a Vespuccian style of exploring temporal distances. Nathalie Cabrol, an astrobiologist based at the renowned SETI institute, exemplifies this trend.

"The Earth" is in fact better understood as a profusion of different planets, Cabrol (2018, 3) proposes that "over geological scales and at low spatio-temporal resolution, Earth may be compared to a multitude of distinct planets, alternatively showing similarities with water worlds recently discovered: Titan or Enceladus." In this particular passage, she stages the ancient Earth as an alien planet, akin to two of Saturn's icy moons. But that is not all: "Similarly, the biological Earth can be reduced to three different planets: the prokaryotic Earth of the Archean, the eukaryotic Earth of the Proterozoic, or the Cambrian Earth with the rise of most major animal phyla" (3). In defiance of conventional biology, Cabrol here interprets key tipping points as *serious* temporal ruptures on a par with the spatial rupture we call the interplanetary medium. This style of reasoning is rife in astrobiology.

We should not be too surprised, the planetary scientist Jan Zalasiewicz (2012, 75, 69) has underlined, by "the profound differences between the alien seas of the Silurian and the familiar oceans of today"—"For it was, then, an alien Earth." When you consider geological eons—the Hadean, the Archean, the Proterozoic, and the Phanerozoic—the Earth has been four different planets, the astrobiologist David Grinspoon insists: "If you came upon it in the different eons you would not characterize it as the same. You would say, oh, that's a different planet."[8] While conducting ethnographic research in astrobiology circles, I encountered such visions of a discontinuous Earth time and again. Astrobiology tends to stress the radically alien quality of ancient Earths. In that sense, astrobiologists are the harbingers of a Vespuccian style of exploration: Temporal distances are taken to be serious boundaries, just as Vespucci imagined the Atlantic Ocean as a serious boundary. Scientists such as Cabrol, Zalasiewicz Zalasiewicz, and Grinspoon effectively introduce an equally worthy complement to outer space: outer time.

The ancient Earth is not portrayed as just a little bit different but as utterly distinctive—as an exoplanet of sorts. The planetary scientist Giada Arney has made the point that research on the habitability of exoplanets has something to gain from taking the "vastly different conditions that have existed in our planet's long habitable history" into account (Arney et al. 2016, 873). The Archean Earth of 3.8 to 2.5 billion years ago, she underlines, was not merely different but *dramatically* different. For one thing, its surface was wholly dominated by single-celled microbes. For another, it was enshrouded in a global, organic-rich haze just as Saturn's moon Titan is today. Just like the latter, the Archean Earth's color was, in all likelihood, not blue but orange—Arney speaks of "the pale orange dot." In the journal *Astrobiology*, she and her colleagues note, "Hazy Archean Earth is the most alien world for which we have geochemical constraints on environmental conditions, providing a useful analog for similar habitable, anoxic exoplanets" (873).In fact, pale orange is just one option in a vast palette of ancient Earths and potentially habitable exoplanets. Besides pale blue dots, scientists nowadays speculate about pale green dots (i.e., "chlorophyl worlds") and pale red dots (suspected to orbit our nearest stellar neighbor, *Proxima Centauri*).[9] "We underestimate," the planetary scientists Joseph Kirschvink told me in a 2014 interview at Caltech, "how chameleon-like our globe is—how relatively easy it is to reach a tipping point after which the entire planet drastically and abruptly shifts into something new and unexpected."

Like Arney, Kirschvink suggests a clean break between our present-day planet and its past incarnations. Clearly, the Columbian style of apprehending the terrestrial past no longer has a monopoly. Scientists like them posit a total discontinuity between then and now, just as Vespucci grasped the existence of a new continent by positing a total discontinuity between America and Europe. Metaphysically, the alienation of the ancient Earth is the equivalent of the Indies becoming America, the outlying islands of a known landmass becoming a new continent in its own right. Planetary scientists such as Arney and Kirschvink effectively rethink geological distance as a serious boundary, just as Vespucci rethought the ocean sea as a serious boundary. The spatial boundary supreme, the interplanetary medium, acquires a temporal equivalent: "Outer space" is complemented by "outer time."

The Columbian style of exploring the ancient Earth has been challenged not only in planetary science but also in another field on which astrobiologists draw a great deal: paleontology. Consider the Earth during the Ediacaran period, between 542 and 635 million years ago. The soft-bodied organisms of that period, known as Vendobionts, remain enigmatic to this day and cannot be connected to modern taxa in a straightforward way. Are they animals, or maybe giant bacteria? Are they perhaps an entirely separate, now extinct kingdom of life? "Alien Beings Here on Earth," one reads in the journal *Science*. Paraphrasing the paleontologist Dolf Seilacher, the author of the *Science* news piece notes, "If we are curious to know what alien forms of life might look like . . . we need not look to distant planets. They exist—or rather existed—right here on planet Earth" (Lewin 1984, 39). Seilacher is a crucial figure in the debate regarding the fossils of the Ediacaran period. He was the first to notice that their basic morphology was radically different from that of the bilaterians, the animals we see all around us today and that have been dominant since the event known as the Cambrian explosion.[10] The body plans of Ediacaran organisms are distinctively flattened: Many of them bear a resemblance to mattresses or quilted pillows. Others have a carpet-like appearance. Some look like fronds, disks, or leaves. They have neither mouth nor anus. What is more, they lack internal tubes. Unlike contemporary animals, they do not possess arteries, veins, lungs, or guts. Essentially, the Vendobionts fed through their outer membrane, by osmosis, and thus behaved like outsized bacteria (see Laflamme et al. 2009). Their anatomical trademark—quilted chambers—in all likelihood served to maximize the body surface available for nutrition and respiration. They were marine creatures that lived in shallow, coastal waters—at the time plants and animals had not yet colonized dry land. The sea floor was blanketed with microbial mats and dotted with pinnacle reefs built by cyanobacteria. The sea itself was oxygen-poor—latter-day fishes would not have survived in it.

The geologist Mark McMenamin (1986) has shown that the Vendobionts symbiotically coexisted with the gooey microbial mats that covered the reefs. They were mat-scratchers, mat-stickers, and mat-grazers. Another remarkable feature that sets them apart from life-as-we-know-it is that their bodies were pervaded with sand: That is why Ediacaran organisms are sometimes called "sand creatures." Their compartmentalized body structure may have been a direct consequence of this internal sand (see Seilacher et al. 2003). Planet Ediacara was a world populated by a veritable "sand menagerie," as McMenamin expressed it. To appreciate the originality of the contributions of people such as Seilacher and McMenamin fully, it is necessary to outline the standard view on the Ediacaran fossils. This view was advanced most authoritatively by the paleontologists Martin Glaessner and Mary Wade (1971). In a fine example of Columbian-style continuity thinking, the latter argued that however strange the Ediacarans may look, they are still our forebears. Glaessner and Wade presumed they had living relatives. That is why they interpreted them as primitive sponges, jellyfishes, annelid worms, arthropods, and suchlike. Seilacher (e.g., Seilacher 1989), however, came up with a counterargument entirely in the spirit of Vespucci. In essence, he suggested that to understand the Ediacaran period, one should not envisage a planet of animals and plants and fungi but, rather, a world utterly distinct: a planet of mattresses and pillows and carpets. McMenamin's vision of a planet of sand creatures is equally Vespuccian in its basic features. The Vendobionts, as the planetary scientist Harry Knoll (2004, 167) has summarized it, "are more alien than animal" in such interpretations.

Crucially, planet Ediacara is not conceived of as just an ancient version of our familiar world. For Seilacher, there exists a total rupture—after all, the Vendobionts disappeared rather suddenly. The point is that he decided to draw a *serious boundary* between now and then, a temporal boundary of the same caliber as the modern interplanetary medium. The quilted body structure of the Vendobionts, he maintained, cannot be considered ancestral to that of the bilaterians, the organisms with internal tubes we are familiar with. Seilacher was at pains to stress that the transition from the Ediacaran to the

Cambrian period was not merely an early stage in the long and continuous diversification of life: "It was the extinction of one highly unusual group of organisms and its complete replacement by another: It was [a major extinction] and it removed a form of life that was alien to all that followed" (Lewin 1984, 39). Ultimately, Seilacher concluded that the Ediacaran Earth was not so much ancient as alien. It was another world, not an ancestral world. If one considers the distribution of landmasses, one astrobiologist once pointed out to me, picturing planet Ediacara as alien is actually rather intuitive: With its continents mainly located in the southern hemisphere and its oceans in the northern hemisphere, it was almost a mirror-image of the Earth we are familiar with. As a celestial body, Ediacara may have been Earth-like, but it was also a very different planet. That Seilacher's ideas have been picked up in astrobiology is not a coincidence. In *Astrobiology Magazine*, Richard Corfield (2009) explicitly introduces a notion of outer time: "In a way, [the Vendobionts] are representatives of an alien world—separated from us not by distance, but rather by time." Later in the same piece, Corfield is even more explicit: "Ediacarans looked like sacks of mud, disks, hubcaps and mattresses—a range of body forms which were never seen again in the fossil record. They were *alien*." Once again, we notice that "outer space" has acquired a temporal complement: "outer time." The interplanetary medium is no longer the only serious boundary thinkable.

Reconfiguring Modern Cosmology: Conclusion

It is no doubt a little maladroit to describe contemporary space exploration in terms of metaphysical styles named after Columbus and Vespucci, two icons of Western colonialism, especially in a volume that seeks to decolonize the space sciences. Yet the predicament of latter-day planetary scientists, astronomers, and astrobiologists is much the same as the metaphysical quandary in which the two Renaissance navigators found themselves: How do I conceive of the boundaries that I am seeking to cross? Are they serious or unserious? In this respect, the basic conceptual options of 21st-century scientists and of 15th-century sailors are essentially the same. What is more, they are unescapable. Since Bruno Latour (2012) published *We Have Never Been Modern*, we know that the modern ideal of a perfectly-purified-science-free-of-any-metaphysical-impurity is and has always been a pie in the sky. To describe Space Age exploration in a more nuanced way, attention to metaphysics, and to styles of scientific reasoning, is critical. Ultimately, this chapter has sketched the contours of an ongoing metaphysical regime change. Modern cosmology can be defined as the conceptual arrangement that operates on the assumption of one serious boundary ("the interplanetary medium") and an infinite number of unserious boundaries. That arrangement may have worked marvelously for a very long time, but it has not always been in place, and it arguably no longer corresponds with the reality of present-day space exploration. In other words, we witness the pluralization of outer space nowadays. As *outer spaces*, planet Mars and planet Ediacara are on a par.

Planetary scientists and astrobiologists have begun to realize that they have options. The status of boundaries does not need to be defined as either serious or unserious in advance. Under the incipient regime, any boundary can be conceived of as either serious or unserious. This means that the number of outer spaces is potentially infinite. This also means that an element of choice becomes inherent to science itself. Some readers will no doubt protest that I exaggerate here. Surely the status of some boundaries has now been settled once and for all? Am I suggesting, in earnest, that America might one day be viewed as a fully fledged alien world again, as Vespucci did? Surely the Atlantic Ocean, or any ocean for that matter, can never become a serious boundary again? Admittedly, there are few signs that a dramatic reversal of the American continent's status is imminent. Yet it is not unthinkable. Consider another continent, Antarctica. It is, as the anthropologist Juan Francisco Salazar (2017a, 2017b) has observed, astrobiology's favorite continent. What is significant is that rather than as a familiar landmass, it is increasingly staged as alien these days. A number of astrobiologists concur with the environmental historian Stephen J. Pyne (2007, 147), according to whom "[the] Antarctic's isolation is so complete that it seems less an intrinsic feature of the Earth than an extraterrestrial presence accidentally slapped onto the planet's surface."

Consider an example: Two planetary scientists I referred to earlier, Michael Carroll and Rosaly Lopes, published *Antarctica: Earth's Own Ice World*. In this book, they recount their voyage to the Mount Erebus volcano and explain how "astrobiologists and aerospace engineers view Antarctica as a proxy for future exploration of oceanic ice moons like Europa and Enceladus" (Lopes and Carroll 2018, 98). In effect, Carroll and Lopes stage Antarctica as a moon in its own right: A geographic continent is rethought in astronomic terms; the Southern Ocean is implicitly reframed as a serious boundary. Depending on the metaphysical settings one prefers, Antarctica can be viewed as the Earth's seventh continent or its second moon—as a terrestrial space or as an outer space. Neither option is superior or "more objective." Current astrobiology shows us that there is more than one way of pursuing objectivity, more than a single legitimate way of enhancing our knowledge about the cosmos. Like the bookish 15th-century navigators portrayed at the outset, contemporary space explorers are profoundly influenced by their "literary preferences": Metaphysical style matters. In summary, the vanilla-only vision of space exploration is no longer fit for purpose: When it comes to researching alien worlds, it is more realistic to think in terms of different epistemic flavors. That is why it is insightful to speak of *styles* of scientific space exploration.

The modern conception of space exploration coincides with the increasingly outmoded scientific ideal of knowing nature bar culture, the cosmos without cosmetics, physics stripped of metaphysics. However successful this characteristically modern ideal of objectivity may have been, it no longer reflects actual scientific practice. In fields such as astrobiology, the very notion of "objectivity" itself has already mutated.

Acknowledging *styles* of space exploration allows for a more realistic account of our current predicament. Any boundary can be configured as either serious or unserious (Tables 3.1 and 3.2). Everything depends on the metaphysical style you, the explorer, decide to adopt.

Table 3.1 Serious and Unserious Boundaries in Space Age Modernity

Cosmology of the Space Age	One Serious Boundary and an Infinite Number of Unserious Boundaries
The Atlantic Ocean	Unserious boundary: Europeans do not view Americans as aliens and vice versa.
The interplanetary medium	Serious boundary: Martian, Ionian, and Europan life forms would be aliens.
Geological and evolutionary distances	Unserious boundaries: The Cretaceous Earth is merely an older version of our current planet. Dinosaurs are ancient rather than alien life forms.
All other boundaries (tectonic fault lines, divisions between biogeographic regions, etc.)	Unserious: They never mark a total discontinuity—that is the exclusive privilege of "outer space."

Table 3.2 Researching Alien Worlds: Metaphysical Options

Researching Alien Worlds Entails Metaphysical Choices	Columbian Metaphysics	Vespuccian Metaphysics
The ocean sea	Unserious boundary: Americans are different from Europeans, but only just a little bit.	Serious boundary: America as a new world, a *Mundus Novus*, rather than as an already known landmass; Antarctica as our second moon
The interplanetary medium	Unserious boundary: Mars, Io, and Europa envisaged as the Earth's cousins	Serious boundary: Mars as an alien dune world, Io as an alien volcano world, and Europa as an alien ocean world
Outer time	Unserious boundary: The Earth, 4.5 billion years of continuity	Serious boundary: The Archean Earth as an alien planet, and the Ediacaran fauna as aliens rather than ancestors

4

Provincializing Earth

Grounding and Writing Acknowledgments from Otherwheres in the Cosmos

David Valentine

> [Sax] crouched and watched the little rodents until he got cold. There were greater creatures out on that plain, and they always stopped him short: deer, elk, moose, bighorn sheep, reindeer, caribou, black bear, grizzly bear—even packs of wolves, like swift gray shadows—and all to Sax like citizens out of a dream, so that every time he spotted even a single creature he felt startled, disconnected, even stunned; it did not seem possible; it was certainly not natural. Yet here they were. And now these little snow pika, happy in their oasis. Not nature, not culture: just Mars.
>
> —Kim Stanley Robinson (1997, 687)

> What's on the earth is in the stars; and what's in the stars is on the earth.
>
> —Stanley Looking Horse, quoted in Ronald Goodman (1992, 31)

> The trip to Mars can only be understood through Black Americans
>
> —Nikki Giovanni (2002, 3)

> What the rockets, shuttles, ships, and landing pods will carry beneath their payload fairing or in their cargo hold, however, along with supplies and satellites, is the capitalist worldview.
>
> —Victor L. Shammas and Tomas B. Holen (2019, 3)

David Valentine, *Provincializing Earth*. In: *Otherwhere Ethnography*. Edited by: Istvan Praet and Perig Pitrou, Oxford University Press. © Oxford University Press (2025). DOI: 10.1093/9780197790885.003.0005

> As hostile as it may seem, the only thing standing between Mars and habitability is the need to develop a certain amount of Red Planet know-how.
>
> —Robert Zubrin (2014)

Before you start,

please note.

This essay is incompatible with Earth. Its arguments are irrelevant to the consequences—to Earth—of the plans of space settlement advocates with all their talk of colonies on Mars, asteroid mining, or lunar property rights. (Elon Musk, Jeff Bezos, and Richard Branson have cornered the market on these visions, but they have many other advocates.) The goal of this essay is not to tell us something about Earth, its history, its nature, or its future, even though these plans are inevitably drawn into urgent discussions of a planet-transforming Anthropocene and the enduring inequalities that animate them. This essay's ethical energy is therefore not supplied by acknowledgment of the frictions of this current terrestrial moment. This essay does not start by orienting itself to the historical formations we name *capitalism* or *socialism* or *social justice* or *colonialism* or *decolonization*, even though they are carried as cargo into the metaphors and rockets, the op-eds and electronic circuits, the visionary plans and critical objections that buzz and crackle through the planning for human communities in other cosmic places. This essay's conclusions are, therefore, not for the sake of any of us—human or nonhuman, biological or geological, past, present, or future—on Earth.

Rather, this essay will ask you to acknowledge, to think, and to feel *from* a speculative, future, Earth-originating settlement grounded on Mars, and to acknowledge such a community as *A Good Thing*. It will ask you to consider where you start when you seek acknowledgment, when you argue over *common ground*, what compatibilities are afforded by the worlds you tentatively share with other Earth life and nonlife, and how the concerns that this essay does not have may appear *provincial* from other places in the cosmos.

I admit, this framing is so alien that it seems it must be a set up for a joke or that it could only be compatible with an unexpected genre in a scholarly collection—for example, science fiction, like the source of the first epigraph above.

But trust me.

It's going to be important that we started our acknowledgments here.

Acknowledgments of Ground

Let's start again.

I begin this materially and ethically ungrounding task by first asking you—since you are likely concerned about the very earthly things this essay is not concerned about—to acknowledge that you and I do indeed dwell on the ground of a planet conventionally called Earth.

I continue by asking you to acknowledge—regardless of differences between us, whatever you call or wherever you dwell on this world or planet, or however you know *for sure* that its nature got to be the way it is—that as fellow Terrans we therefore share common, enduring, grounding conditions because of our planet-world-dwelling on Earth, conditions that are unique in the universe.

But hold on.

I suspect you have objections. Anthropologists routinely dig into claims such as this—about common *humanness*, *nature*, *world*, and *universe*—to reveal their epistemological and ontological grounds and cargo. Our critiques reveal the frictions and incompatibilities of power, privilege, violence, and stratification that appeals to commonality both presume and ignore. Thinking about the second request, then, indicates that the first—simply to acknowledge that you and I are both *on Earth*—may be hiding something.

But please, throw me a bone.

If possible, I ask that you *trust me* just for a bit longer, that you defer your objections simply for one page. I promise I will get to them.

Here are only a few of the entangled, grounding, earthly conditions—note that I explicitly do not call them *nature*—that I ask you to acknowledge as *common*, *enduring*, and *unique*: a dense 78% nitrogen/21% oxygen atmosphere with a sea-level pressure of 1 atm, that, along with Earth's magnetic-field-producing iron core and its approximately 150 million km distance from the sun, protects all life on Earth from deadly solar and cosmic radiation (Canfield 2015); a 9.8 m/s^2 gravity field within which terrestrial entities have habituated to motion and rest, and which allow material processes (*acceleration*, *inertia*, *friction*) to unfold so that we may crouch, wait, suffer, or carry cargo in varied but anticipable ways (Gordillo 2020; Valentine 2017a); a 24-hour rotation on an axis tilted 23.439° relative to the plane of Earth's 365.25-day orbit around a middle-aged, unusually quiescent G-Type main sequence star that, in combination, confer a day/night cycle and annual seasonality; and billions-of-years-deep ecological-geological histories that enable us all to be in compatible chemical and metabolic relationships with diverse forms of terrestrial plant and animal life (Fortey 2005) and other inanimate entities like *ground* or *soil* (Salazar et al. 2020). I base my arguments on the claim—and still, I must ask you to *trust me* for just one more sentence—that whoever or wherever you are or whatever you know about how this planet-world and all of us got here, we share these conditions as the common grounds of experience and futurity.

But wait.

I should get to the objections I have so far deferred. In subsuming all human worlds into a commonly shared and equally distributed ground with the unmarked plural pronouns *we* and *our*; in asking you to acknowledge Earth's conditions—in English, in the idiom of modern Western science—as outside historical time; and in asking you to trust me—even if just for a bit—I must start again with a different acknowledgment of ground.

And so.

I acknowledge in historical time (specifically, 05h16m:27s CST or 9h:16m:27s UTC on March 16, 2024 CE) that I sit and write in a specific place—the settlement of Somerset in the region currently known as the U.S. American state of Wisconsin—in/on a planet-world and continent that only some of us call *Earth* and *America*. This ground is on the traditional land of the Dakota people, stolen by force and imposed law by European settlers as the grounds of U.S. sovereign power (Nichols 2013), a process my presence here continues. It has been partitioned, cultivated, and excavated without acknowledgment of the friction arising from prior and ongoing Indigenous worlds and sovereign claims. And—like this essay—this land-taking has been grounded in settler claims of commonality, scientific rationality, and requests for *trust*, despite the repeatedly broken promises of settler states. These states treat decolonizing demands for territorial sovereignty as—alternately—jokes or threats, even as states demand that Indigenous peoples *wait* for their full acknowledgment as future citizens in settler terms (A. Simpson 2014; Rifkin 2017; Tuck and Yang 2012). In turn, Indigenous theories open to radically different acknowledgments of where Earth came from and how it came to be, and of the mutuality of relations among human and nonhuman entities on Earth (L. Simpson 2011) and in the cosmos, evident in the Lakota star knowledge offered by Stanley Looking Horse in the second epigraph above (Goodman 1992). Or Maoli Kānaka (Native Hawaiian) and other Polynesian wayfinding traditions may be acknowledged as Indigenous ways of exploring the cosmos without the presumption of property rights and extraction (Chang 2016; Hau'Ofa 1994).

Other historically grounded acknowledgments proliferate from querying commonality as the starting place for acknowledgments in space or time (Valentine and Hassoun 2019). Primary among them are the enduring consequences of the brutal trade of West Africans as cargo into American chattel slavery (Gumbs 2018; Sharpe 2016) and the foundational exclusion of Black lives from the Enlightenment constitution of *the human* (Spillers 1987; Weheliye 2014). As such, Afrofuturist fiction, scholarship, and poetry may start with the apocalypse of middle passage as a resource for thinking for how to embrace the challenge of a one-way trip to Mars, as poet Nikki Giovanni (2002) does in the third epigraph above. Likewise, acknowledging the ableism in universalizing descriptions of human bodily practices—my *writing* or *sitting* on this land or the *carrying of cargo* without consideration of my embodied capacities (Kafer 2013)—may reveal different imaginations of (dis)ability in space (Wells-Jensen et al. 2019).

Intersecting critical approaches—feminist, queer, subaltern—similarly demand attention to frictions arising from an assumed common global dwelling (e.g., Spivak 2005). Intellectual-political engagements with ontology (Viveiros de Castro 2014), matter and objects (Behar 2016; Bennett 2010; Harman 2018), and human/nonhuman imbrications on an Anthropocene planet (Latour 2019) reveal the frictions in post-Enlightenment thinking about a *common world* by emphasizing the capacity for *worlding* by nonhuman and inanimate entities that will not wait for the conclusion of human projects (de la Cadena 2015; Kohn 2013). These investigations seek out *otherwise* earthly worlds (Povinelli 2016; Harvey 2000; Appadurai 2013), challenging the hold of the Western epistemological tradition—and its co-traveling cargo of capitalism—on the future. Scholars have further shown how *Earth* itself—and the very planet-ness of Earth my opening request turns on—is an achievement of Enlightenment philosophy, scientific modernity, globalized capitalism, and Cold War militarization (Cosgrove 2003; DeLoughrey 2014; Lazier 2011), even as others argue that Earth's planetary materiality demands critical attention (Clark and Szerszynski 2021). *Terraforming*, from this perspective, might best describe Earth itself under Anthropocene transformation (Haraway 2016; Woods 2014) rather than fantasized, science fictional re-makings of other world-planets.

And so.

Space settlement visions of colonization and extraction are thus easily excavated to reveal the frictions in their ideological and historical cargo.

While they draw examples mostly from near-Earth, contemporary human projects in space, critical scholars of outer space argue that future space settlement will clarify and extend existing terrestrial inequalities to the cosmos (MacDonald 2007; Dickens and Ormrod 2007). They thus draw attention to the unacknowledged ideologies that underpin these plans, as Shammas and Holen (2019) make explicit in the fourth epigraph above. Concomitantly, critical scholars concerned with outer space activity argue that such plans draw attention away from real, immediate matters of terrestrial concern (Klinger 2019). These scholars urge *us*, thus, to *provincialize outer space* (Redfield 2002) and to come back *down to Earth* (Latour 2018) to attend to the devastation of capitalist and enduring colonial relations on Earth. Alternately, they provoke dismissal as ungrounded fantasy, as just science fiction (Ormrod 2009).

In short—regardless of where they start—terrestrial scholars, artists, poets, and activists have deeply grounded toolkits to dig into any unmarked claim to a shared *humanness*, a common *world* or *nature* or *future*, or a neutrally encompassing *Earth* within a containing *universe*. In this light, the prospect of future, market-driven "benefits to all mankind" arising from "making humanity a multiplanetary species" (as Musk frequently says) seems materially and ethically ungrounded from Earth's multiple crises, even as space settlement advocates ask *us*—as I have asked you—to *trust them*. And so, it absolutely matters which of these grounded acknowledgments you start with when you seek to explain something about the consequences of human settlement of the cosmos for the world-planet that you and I tentatively share.

But wait.

Please remember.

This essay is not *about* or *for* the many worlds of Earth.

Rather, it is *about* and *for* a not-yet Martian, terrestrial-originating world (who- or whatever they are, however they got there, whatever they think about Earth). It asks about the consequences, *for them*, of what may be carried as epistemological and ontological cargo from Earth to cosmic otherwheres if the acknowledgment of Earth's grounded, material human histories always takes precedence over the acknowledgment of Earth's grounding material conditions in the founding of those histories. The cargo I am concerned with in this chapter, then, is not so much ideological—whether settler or critical or Indigenous in origin—as it is conditional. That is, I seek to dig into how Earth's arrangement of material conditions implicitly grounds the outlines of common arguments between space settlement advocates and their critics about human presence in other cosmic places *without acknowledgment of those places' material arrangements.* Indeed, the conditions I invoke—gravity, solar distance, axial tilt, atmospheric pressure, magnetic field, planetary core, cosmic and solar radiation—are rarely present in terrestrial arguments over what will become of *nature*, *humanness*, *world*, or what is *universal* after humans leave Earth. Both settlement planners and critiques of them commonly presume that after the technical establishment of a rudimentary, Earth-compatible ecological bubble, human actors can go about the business of extending terrestrial historical formations (*capitalism*, *socialism*, *gift economies*, *settlement*) through common analytics (*politics*, *ethics*, *networks*, *governance*, *decolonization*, *systems*, *exchange*). But this shared presumption fails to acknowledge that mundane—in the sense of both ordinary and terrestrial—forces, qualities, and actions (*friction*, *crouching*, *failure*, *resistance*, *externality*, *digging*) underpinning such formations and analytics depend on planetary-scaled arrangements of material conditions that cannot travel beyond Earth as cargo. Crucially, Earth's common, settled conditions are precisely what allow me to ask you to *wait*, to *defer*, and to presume—even as I acknowledge all the terrestrial histories outlined above—that your deferral across the span of reading a few pages of text is not *inherently* an immediate threat to your life and to the persistence of all life and inanimate assemblages in your world.

But in other cosmic places—any other place in the universe that is not Earth—it could be. *No. It would be.*

In the world I will write of—a particular place several meters under the equatorial plane dubbed *Elysium Planitia* on the planet-world still called Mars by many Terrans—the request to *wait* or *defer* in relation to an apparently *mundane* reading practice presumes that *you* or *we* are not responsible for maintaining the conditions of worlding itself: a nominal ratio of oxygen to an inert gas in the immediate environment; the protection of all living things, machines, or entities from cosmic or solar radiation; or the amplification of the sun's energies to keep plants growing. In this place, my *writing* and your *reading* would need to be explicitly acknowledged and made compatible with the maintenance and continuation of a whole world. And vitally, the actants indexed by *you* or *we* would not necessarily be *human* by any standard we interrogate in this world, perhaps not even what any of *us* would call *alive*. As such I will dig into what happens to the grounding of *nature*, *humanness*, *world*, or *universe* if Earth—wherever you are in/on it, whatever you call it, whatever your experience on it, whatever you know to be true about its origins and histories, and what is possible about its futures—does not provide the conditioning of the ground that you/we start from.

I depend on the groundbreaking work of anthropologists of outer space who ask us to consider *space as itself* in low Earth orbit (LEO) (Battaglia 2012), who show how human presence in near-Earth space transforms concepts of *environment* and *system* on and off Earth (Olson 2018), or who show how cosmic sites are imagined through *place-making* practices on Earth (Messeri 2016). But I will ask what becomes of these formations or places and what *space as itself* demands when humans are 979 times further away from Earth than its moon, or more than a half million times further away than LEO where Earth's conditions or metaphors cannot be either a resource or an escape plan. Likewise, while I acknowledge with gratitude the energy I draw from Indigenous, Black, disability, feminist, queer, and other distaff terrestrial intellectual traditions in their demand to consider where one *starts* an account of the world, I argue that they are also not enough, by themselves, for the project I am launching here.

As such, I jump over—a metaphor grounded by 9.8 m/s^2 of gravity—the usual focus of critical accounts of outer space activity, bound as they are to Earth and its concerns, to make an ethical claim for, and take the intellectual step of, casting my lot with the lives and beings of a not-yet thriving Mars community peopled by generations of Earth-originating human and—significantly—other entities. I therefore need to accept the claims of space settlement advocates such as Robert Zubrin, quoted in the fifth epigraph above, that the initial establishment of human settlements in other cosmic places is possible using current technology.

And you may think: *Oh, this is just science fiction.*

But even if it is, this essay will demonstrate that the extension of Earth's histories—whatever you think of them, wherever you stand-sit-crouch-hunch-in-pain in relation to them—to otherwheres in the cosmos is, materially, the most science fictional scenario of all. I will speculate in an orthogonal decolonizing register on the different kinds of—and transformations of—frictions that emerge in starting a conversation from a place that is indeed not yet colonized, but also a place where none of Earth's deferring conditions can be carried as cargo. (This will require a Martian retooling of Tsing's [2005] account of *friction*). As such, *decolonization* here indexes the perspective of a different, Martian *you* who might indeed understand *Earth* as a single, common—and alien—place.

I thus seek a common frame—for you and me—not to erase or explain differences among us *here* but to *provincialize Earth* and its materially conditioned histories to make the case that they are incompatible grounds for acknowledging what may unfold should humans establish worlds in cosmic otherwheres. (This will involve a Martian recalibration of Chakrabarty's *Provincializing Europe* [2007] or Redfield's call to *provincialize outer space* [2002]). I further intend to make the case that there is (what we on Earth may refine out as) an *ethical* reason to do so, not *for* any of *us on Earth* but *for* and *from* the grounds of cosmic otherwheres.

(*You may ask: What does Mars have to say about this? Please wait, we'll get to that.*)

Drawing on a decade of research among space settlement advocates, on encounters with my professional peers who weigh in on human cosmic futures, and on a certain science fictional know-how, I will argue that both space settlement visions *and* critical analyses of them consistently bury the material differences between Earth and the multiple otherwheres of outer space. These differences demand a reading of *decolonization* in a different register and can be figured through "ground" in two terrestrial senses: (1) ontologically as a material entity specific to a cosmic place at a condition-setting scale; and (2) as a terrestrial metaphor for epistemological and ontological presumption of enduring background conditions for *human* explanations of and living-with *nature*, *world*, or *universe*. Focusing on a specific place under the surface of Mars, I argue that space settlement plans, while they must certainly be subject to terrestrially grounded critical questions for the sake of a tentatively shared Earth, open to different grounds for imagining relationality and groundedness beyond Earth—including the capacity to hold the ontological and metaphorical separate. Remember: This essay is not about the many worlds of Earth. *Its conclusions are not for you.*

But wait.

(*You still can. Waiting-to-read-on* in itself *won't kill you and end your whole world. Not* right now.)

There may yet be something in here for me and for you, those who remain on Earth, even if it is only a residuum. And so, residually, I will also argue that an anthropology of outer space that only digs into the ideological or historical grounds underpinning space settlement visions refuses and defers the materiality of other cosmic sites as grounds from which to develop radically different accounts of nature, humanness, world, or universe.

But trust me.

This is genuinely a minor, peripheral, *provincial* possibility. It is *not* the goal of this essay; if it were, it would remain grounded here, on Earth.

So, let's go.

Let's dig into these outlandish claims by looking at them *from* the ground of Mars.

A Moose and a Mole on Mars (or: Geology Versus Areology)

Kim Stanley Robinson's sprawling *Mars* science-fiction trilogy—from which this chapter's first epigraph is taken—unwinds over a near-future 200-year terraforming of Mars. In *Blue Mars* (1997), Saxifrage "Sax" Russell, a physicist, is astonished as he crouches on Mars's surface, observing the unnatural-uncultural fact of moose, snow pika, and other Earth-originating animals wandering its surface, protected by a human-initiated atmosphere. But centuries before, in the trilogy's first novel, *Red Mars* (Robinson 1993), the First Hundred terrestrial scientist-settler-arrivants (Byrd 2011) dig down to establish their vaulted habitat several meters underground, in acknowledgment of Mars's current deadly-to-Earth-life surface conditions—in particular, its radiation profile.

But wait.

Digging into Mars is not just science fiction. It has been attempted, starting at a time coinciding with the common Earth time/date 19h:52m:59s UTC on November 26, 2018CE when the joint National Aeronautics and Space Administration (NASA) and European Space Agency (ESA) InSight lander settled down to its work of digging a hole at a point on Mars designated from Earth as 4.5°N 135.9°E on the vast equatorial plane that carries the Earth name *Elysium Planitia.*

InSight is one of many probes, orbiters, landers, and rovers that, since NASA's Mariner 4 in 1965, have taught us Terrans much about Mars' ground in a scientific idiom and that—especially in returned images of Earth-like deserts—makes it appear tantalizingly similar to Earth's (Messeri 2016). Mars orbits the sun in the same elliptical plane as Earth, rotates on a similar axial tilt (25.19° to Earth's 23.439°) that provides annual seasons, in a period—or sol—remarkably close to Earth's day (24h:39m:35.244s by the common terrestrial clock).

But wait.

Mars' ground is conditioned by different arrangements than Earth's, putting their apparent compatibilities and respective regularities out of step. For example, Mars' distance from Earth varies between 54 and 400 million km, but on average, its orbit is one and a half times further from the sun than Earth's, receiving 43% of the solar energy Earth does. Its relatively eccentric orbit makes its furthest and closest approach to the sun more extreme than most other planets, producing seasonally extreme temperatures. The resultant 669 Martian-sol solar year (687 Earth-days) is only 40 Earth days shorter than 2 Earth years. For these reasons, it is only possible to launch a mission to Mars once about every 26 Earth months.

Mars is smaller and less dense than Earth, producing a gravity field—3.71 m/s^2—about a third of Earth's 9.8 m/s^2, and unlike Earth, Mars likely has a dormant core and thus no global magnetic field. These factors likely caused the loss of Mars' primordial atmosphere to space and the absence of any obvious extant life. The remaining, mostly carbon dioxide, atmosphere is 0.61% the density of Earth's—extremely thin by comparison. This means that its surface is bombarded by both solar and cosmic radiation and by meteorites, dangers that are mitigated on Earth by its dense atmosphere and magnetic field.

Mars also seems to have been warm, wet, and lively long before Earth, but its extant water is frozen within its rock or regolith because of surface temperatures as low as −225°F (−133°C). Furthermore, its ground is covered by finely grained iron oxide material containing pervasive and high concentrations of perchlorate, a toxic salt much rarer on Earth even as, unlike on Earth, there is relatively little nitrogen, an element essential to terrestrial plant growth. (These factors make *soil* a poor synonym for this *ground*). In short, Mars' ground would be incompatible with terrestrial life, which is why planners and builders of Martian settlements—science fictional or not—dig down, reassigning Earth's atmospheric and magnetic field protection to regolith: to ground.

But *we* still know little about Mars' underground. InSight is investigating Mars' seismology, heat flow, and deep interior structure, and one of its first discoveries was that Mars is subject to significant seismic activity that would need to be acknowledged by any future settlement architecture. Yet, some of the research team's questions have been deferred because of unexpected interactions between Mars' conditions and terrestrial presumptions about *ground*, *soil*, and *geology*. InSight carried a Mole as cargo to Mars, not an Earth-originating animal—such as those observed by Sax in the first epigraph above—but an instrument designed to dig 5 m into Mars' ground to explore its *geological history*. However, after digging down 30 cm, it stopped. According to the project's principal investigator, Tilman Spohn, the problem lay in the intersecting conditions of Mars itself: "'We are now rather sure that the insufficient grip from the soil around the Mole is a problem, because the friction caused by the surrounding regolith under the lower gravitational attraction on Mars is much weaker than we expected'" (Gough 2019). Meanwhile (as we say on Earth), Mole presented other puzzles for Terrans, such as the fact that its digging never produced a lip of material—or *ground* or *soil*—around its rim as it would on Earth.

In late 2019, the team jury-rigged InSight's *soil scoop* to push against Mole, increasing the friction. But even more inexplicably, while initially successful, Mole was subsequently ejected from its hole. In early 2020, CNN reported that the team had employed the scoop again, joking that "NASA fixes Mars lander by hitting it with a shovel," a "charmingly rudimentary" approach to a high-tech problem (Guy 2020). But in the end, on Earth date January 14, 2021, the research team gave up and acknowledged that they had to defer their questions. "We've given it everything we've got, but Mars and our heroic Mole remain incompatible," Spohn said (Good et al. 2021).

Meanwhile (as we keep saying on Earth), terrestrial robotic craft steadily return inexplicable findings about Mars, such as methane emissions from unknown sources (ESA 2019) or unusual fluctuations in local magnetic fields (Howell 2020). That is, despite Terran scientists' reasonable presumption of the universality of rudimentary conditions and forces such as *chemistry*, *friction*, or *gravity* or the apparently simple technical act of *digging a hole* in other cosmic places, those phenomena are grounded by arrangements among them that produce proliferating differences not fully subject to acknowledgment, in advance, from Earth's settled surface. Or, to start from Mars, we might say that Mars' ground will not defer its own conditions to acknowledge the grounding presumptions of earthly *soil*, the terrestrial presumptions of *geo-* in *geology*, or the workings of terrestrial machinery in its conditional arrangements. Well in advance of moose roaming terraformed Mars, areology will not defer to geologically conditioned inquiry.

Cargo from Earth

So now,

we can dig into the frictions arising from the presumed-in-advance compatibility of Mole and Mars to explore the possibility that other kinds of material processes—physical or historical—may also not travel easily as cargo from Earth's deferring ground.

Robinson's scientific realism is attractive to space settlement advocates whose plans for Martian habitats also tend to dig down. However, the commonality in science fictional and settlement planners' Mars habitats lies not only in technical solutions but also in a shared conviction about their compatibility with Mars' conditions. Robert Zubrin, the preeminent advocate for near-term Mars colonization, argues that the permanent settlement of Mars is not only possible with existing technologies but also essential for "reaffirming the pioneering character of our society" (2020, xxxiii). But like Robinson or the InSight team, Zubrin also relies on Mars' commonality with Earth in the presumed universality of rudimentary, material processes and resources as they are arranged on Earth, and the resulting ability of pioneers to jury-rig them to make Earth-compatible settlements. Unlike some advocates, Zubrin does not consider subsurface radiation protection to be essential, although he plots out one vault-like structure recalling the settlement constructed by Robinson's First Hundred (189). But regardless of disagreements over technical matters, settlement advocates rally around Zubrin's claim that all terrestrial settlers need is a "certain Red Planet know-how," a rudimentary technical alignment, to establish the conditions for personal and entrepreneurial freedoms in a pioneering Martian colony, unconstrained by Earth's histories.

But wait,

because critical scholars, such as David Harvey (2000) and Frederic Jameson (2007), have also dug into the *Mars* trilogy, not for its science but for what they see as its realist extension of material historical processes to a future, peopled cosmos. In contrast to Zubrin, Harvey (2000) finds compatibility in the plotting of terrestrial histories of exploitation and stratification on Mars and embraces Robinson's historical know-how in narrating "innumerable cross-references to the actual historical geography of imperial conquest and colonial and neocolonial activity as promoted throughout the long history of capitalism" (191). Like Harvey, Jameson finds the

Mars trilogy—and its emergent gift economy among settlers—instructive for thinking through *utopia* as a political possibility. As with space advocates, opinions among their critics are mixed. For example, Shammas and Holen (2019) see only the possibility of an extractive *capitalist worldview* being carried as cargo beyond Earth. But regardless of disagreements, critical scholars' historical–materialist know-how unites them in the conviction that Mars opens to the continuation of terrestrial historical materialism.

But wait.

This comparison is not intended to reveal the ideological roots of Zubrin's bootstrapping, exceptionalist vision via Harvey's or others' critical excavations. Nor is it to adjudicate the alternative technical or historical futures that settler advocates and their critics argue over. For, from Mars' ground, what is most striking is the compatibility *across* these apparently competing accounts: the common and implicit agreement among settlement advocates and their critics that after the technical establishment of a rudimentary, Earth-life supporting ecological bubble, terrestrial historical processes—*settlement*, *exploitation*, *market forces*, *pioneering*, *stratification*—can and will unfold much as they did on Earth through human action—*digging*, *watching*, *crouching*, *waiting*—unimpeded by Mars' conditions. And in both, *capitalism* or, alternatively, a *gift economy* or *socialism* are the only options available to carry as cargo to Mars's surface.

But what might Mars—or rather, a particular, humanized place just under the surface of Mars—have to say about these traveling formations? As we will see, Robinson's acknowledgments of Martian difference are not enough for they also defer Mars's grounding conditions in their binding of Mars to Earthly historical and conditional materialities.

Mars Won't Wait

Deferral—and the question of where to start an acknowledgment of ground—emerged ethnographically at the professional space conferences, spaceports, teleconferences, and online forums where I conducted (what we call on Earth) *fieldwork*. In these sites, I spent long terrestrial hours with space settlement advocates who engage in techno-social speculation about the compatibility of terrestrial phenomena with Mars, the moons of the outer planets, or even in massive, rotating artificial-gravity colonies.

Perhaps surprisingly, then, the majority of talk I heard was about the prior problem of how to leave Earth in the first place. Once in LEO, the additional energetic and technical demands of going elsewhere in the solar system are relatively low. Achieving LEO, on the other hand, is not only difficult but also expensive, and so developing a *low-cost, reliable access to space* system (LCRATS) is vital for settlement plans. Regardless of political affiliation among participants (and it is broader than most critical commentators imagine), settlement advocates are almost universally in agreement that free market terrestrial arrangements are essential for establishing LCRATS. At the same time, the significant costs and logistics of space infrastructure demand huge public investment, and so a key concern among advocates is how to prevent concomitant overregulation of entrepreneurial space activity. It is, therefore, in leaving Earth that Earth's material conditions are essential to acknowledge, and so their binding to Earth's material histories and then to the rest of the cosmos by both settlement advocates and their critics—if with different valences—is unsurprising.

But in other conference sessions, scientists, architects, doctors, lawyers, and engineers worked through the other side of the equation: *What would it be like* on Mars, in an artificially constructed, cylindrical space settlement, inside a main-belt asteroid, or on a moon of Saturn or Jupiter? Here, the compatibilities of the terrestrial to the cosmic felt less secure. Like scientists—such as the InSight team—remotely exploring Mars or exoplanets (Messeri 2016; Vertesi 2015), space settlement advocates use their Earthbound bodies, metaphors, and expertise to plan for alien forms of world-difference, but unlike those scientists, settlement planners must consider human presence on alien grounds. Conference sessions covered radically different topics: a proposal for a Mars transportation infrastructure, three-dimensional tool printing in reduced gravity, the pitch of a staircase in a rotating habitat, the chemistry of concrete walls on Mars, radiation protection in a colonial transport ship via an accreting wall of human waste, and so on. The question and answer periods after such presentations burst with equally varied suggestions, alternative interpretations, skepticism, or even downright disagreement from audience members speaking from their own areas of expertise about the consequences to human life and habitats arising from incompatibilities with unexpected forces or entities.

As with my requests of you to *wait*, these interactions also allow deferral, as they are subtended not by the alien worlds speakers reach out to, but by homely, terrestrial conditions. And so, disagreements over what comes next, what interactions count, or what consequences might emerge (from unexpected heat flows, unanticipated chemical transformations, potential behavioral changes among human participants, novel intersections of terrestrial technology with alien places, etc.) can be deferred. At the next conference—here on Earth—or the one after, new data, new experiments, new alignments might open to new compatibilities.

But Mars won't wait.

The fundamental problem that unites technical settlement visions, critical accounts of them, and metropolitan science fictions is that the human action presumed by them also presumes the persistence of stable, deferring conditions within which that action can unfold. In an imagined future settlement under the Elysium Planitia at 4.5°N 135.9°E—let's call it *Giovanni Prime*—the technical substitution of Earth's atmospheric and magnetic-field radiation protection with regolith only *multiplies* questions of compatibility. Architects planning concrete walls must acknowledge not only Mars-specific available resources and chemistry for appropriate binders, curing, weathering, and load-bearing in 3.71 m/s^2 at 0.006 atm atmospheric pressure but also their susceptibility to water flows that might emerge from subsequent release of regolith-locked water due to heating from a settlement (Larson et al. 2012). Simultaneously, within a world tentatively achieved via locally made concrete, labor law, farming, and waste would need to incorporate relations among physics, potential human conflict, energy needs, and unanticipated interactions with microbes or Mars' persistent chemistry (Valentine 2017). Moreover, for a sustainable settlement, terrestrial histories of race and gender would need to be reformulated into the day-to-day of crew composition, genetic diversity, reproduction, and ethical practice (Smith and Davies 2012). But the punctuated possibilities of Earthly writing—gathering, as I have, compatible terrestrial domains within discrete sentences—fail to acknowledge that all these elements–forces–concepts–things–actors are in persistent and immediate relation to one another in a *particular* place—not the earthly *no place* of utopia—and cannot be kept apart. (And it fails to acknowledge the time, date, or location in which these words were written because these historically and conditionally material realities are not essential—on Earth—for following this argument and its consequences.)

Crucially, *any* failure, relation, mistake, or unforeseen problem in *any* of those relations at *any* scale could end the *entire world* for everyone and everything within its boundaries *right now* because of Mars' incompatibilities with terrestrial forms of life. And so, human action and intention would only be one of the proliferating potential agents in play. An unexpected flow of water, a technological breakdown, an out-of-the-ordinary carbon cycle, or a genetic mutation could equivalently end an entire world. While the time frame of these ongoing world-concerns may emerge over days, months, or years, there is the always-potential for them to erupt immediately, an existential, *universally* world-ending event for every*one* and every*thing*.

And another thing:

As on Earth, Mars' conditions facilitate variable natures across its ground, so *Martian settlements* in different places would demand different responsiveness. (Mole's failure arose in part from terrestrial scientists' assumptions about the global commonality of Martian *soil*.) Concomitantly, Mars is not Ganymede (a moon of Jupiter) or Titan (a moon of Saturn) or Ceres (a main asteroid-belt dwarf planet), potential locations for human settlements that are as conditionally different from one another as each is from Earth. These places' commonality lies, therefore, not in their openness to terrestrial histories and human action but, rather, in their conditional challenges to terrestrial—-and to each other's—orders.

Terrestrial theoretical approaches that counter the human-centering of the Western Enlightenment–colonial tradition by emphasizing the agency of nonhuman actants are not enough to account for this alien relationality. Such perspectives merely shift the stabilizing work of *the human* in modernity to the unique conditions of Earth itself (see Clark 2010). These knowledge-projects still depend on terrestrial distinctions that keep terrestrial relationality in place but that are incompatible with nonterrestrial places: between labor law and Earth's specific chemistry; between history and the quantity and quality of radiation received on Earth's surface from the sun and the cosmos; between ethics and specifically earthbound forms of terrestrial microbiology; or between kinship and 9.8 m/s^2 of gravity. Off-Earth, each domain is simultaneously found wanting and demands acknowledgment—without deferral, priority, or ontological or metaphorical security. Moreover, the apparent flatness of domain-crossing relations in an off-Earth world would be shot through with eruptive peaks, for any *individual* entity or actant could also end the entire world for every*one* and every*thing*, demanding a thorough respect for each thing's/one's autonomy and abilities at any scale. Off-Earth, neither "the human" nor any other entity, force, or agent could occupy a privileged or ontologically secure status; but neither can terrestrially imagined *relationality*, *networks*, or *systems* (Olson 2018). This difference is not simply one of content that requires revised technical or historical know-how, but rather demands of "relation" a prior acknowledgment of the absence of terrestrial groundings. It is for this reason that I argue that the *ontological* and the *metaphorical* are also exclusively terrestrial domains, for to know or metaphorize a world presumes its ontological continuation without concern that an ill-fitting metaphor may bring it to an end.

Above all (as we say on Earth), there is no externality to escape to, for outside—*above*—this place is precisely what this place-world is designed to exclude to the fullest extent possible. Earth's deferring capacities may allow a material or metaphorical *breathing-space, solid ground*, or a resetting *outside* in a crisis. (This is why *I can't breathe, stolen land*, and *racialized incarceration* index specifically terrestrial scandals.) But they cannot travel as cargo to Mars so that humans can go about the business of being on-Earth-on-Mars.

Because Mars won't wait.

Sax's *crouching* on Mars condenses the terrestrial commonalities among metropolitan science fictions, settler planning, and materialist critiques of them. Like this essay, these forms of knowing-Mars-from-Earth *jump over* spacetime. But unlike this essay, they do so to a spacetime when Mars is always already secured for human action focused on Earth-sourced projects without ongoing attention to the immanence of this specific world's continuing- and ending-possibilities. These accounts accentuate the differences of alien places when they are significant plot points in terrestrial narratives, fictional or not. But to make alien-ness compatible with terrestrial matters of concern, they must defer alien conditions at the points that Earth-mundane human actions (*writing, sitting, waiting, crouching, breathing, reading*) and the diverse terrestrial-human projects upon which they depend (*colonization, social justice, exploration, decolonization, extraction, diversity*) must align with the provincial limits and consequences of terrestrial historicity.

This spacetime jump further presumes the endurance of human-terrestrial concerns over the spacetime of settlement itself, uninflected by Martian accommodations to what *nature* might have become in the 26 months since the last spacecraft arrived from Earth, new compatibilities with *world* innovated in the up to 40-minute communications lag between Earth and Mars, revised Martian theories of *humanness* in relation to lively concrete and artificial intelligence (AI) and microbes, and what might be *universal* if you started from a place where *the universe* does not appear accessible to you from a simple step *outside*. In short, Sax's ability to *crouch* and ponder the earthly question of Mars' *nature* or *culture* stands in for almost all terrestrial space projects, settler or decolonizing, scientific or fictional: What is possible or relevant on Mars, from Earth, must be secured by a terrestrial conditioning that stabilizes the differences of Mars from Earth's ground.

This materialist—rather than science fictional—insight challenges, in advance of a colonial project, the terrestrial presumptions of both space settlement advocacy and its critics who see outer space settlement either as a solution to the earthly problems or as a ruse to extend terrestrial inequalities off Earth's surface. The unacknowledged cargo in these futurisms is not a particular *ideology* but, rather, the presumption that compatible frames can be established *within which* terrestrial action or analysis can unfold after a brief wait for technical alignment—a certain know-how, or a certainty of capitalism's or humans' universal technological reach—to accommodate pre-established, terrestrial forms of difference and history.

And still, Mars won't wait.

Science F(r)iction from Mars

And you might say . . .

. . . But terrestrial technologies *have* successfully demonstrated compatibilities between Earth and Mars! A helicopter has even flown in Mars's thin atmosphere! Surely, these technical successes are merely preparatory of capital and colonialism's triumphant arrival on Mars?

And you might say . . .

. . . But *we* are all still here *on Earth*, burdened with legacies or armed with privileges that whirl through the talk about Mars colonization. How can any materially grounded account not start with acknowledgment of this cargo? And what justifies presuming an Earth-originating Martian settlement as *A Good Thing* when Mars has its own orbit, its own past and future?

And you might say . . .

. . . Even if this is an accurate account of Martian difference, then is not the very possibility of a Mars colony revealed as a joke, a privileged, post-Enlightenment, White-guy science fiction?

But wait.

(Please recall that this essay's multiple requests for your deferral and your patience draw attention to the enduring possibility, conferred by Earth, for me to write, for you to respond and—*meanwhile*—for us to struggle with the ethical and material demands of daily life, subtended by Earth's enduring conditions. There is no *inherent* world-ending potential to your reading, right here-now.)

And remember, while you wait,

that technical successes are those that *make Mars seem compatible to Earth.* They obscure much more common failures—like Mole's—that reveal the back-and-forth second tries and jury-riggings constantly demanded by Mars but conducted from Earth's stable ground—like the debates between experts at a space conference or the exchange I imagine between you and me. However, the presence of humans alongside other terrestrial entities on Mars both transforms relations and introduces a relationality you cannot choose, defer, or retry. *Mars won't wait* to acknowledge the exceptional human achievement of a first colony—as if Earth's conditions went along for the ride as cargo—without that colony's ongoing acknowledgment of Mars' ground *as itself* (Battaglia 2012).

And we certainly must not forget, as we debate,

that all the talk of Mars colonies as *destiny* for *all mankind* or as a *settlement* of past crimes are not enough for the humans and nonhuman entities suffering on Earth *right now* under the weight—materially within Earth's 9.8 m/s^2 of gravity—of inequality or violence, of stolen land or slavery or genocide. But also remember that terrestrial suffering—or responses to it—will not end all worlds for everyone and everything else on Earth *right now.* In other words, Earth's enveloping conditions nurture its inequalities as much as its abundant gifts and deference to the future. But on Mars all *a slave* or *prisoner* or *dissident* would need to do is dig a hole in the wall or defecate in the wrong place or persistently use an ill-fitting metaphor to demand her freedom—briefly, before the end of the world. (Presuming the stable source domains of the earthly metaphors I've used in this essay—*corner the market*; *the xth epigraph above*; *throw me a bone*—could be as materially deadly as misplaced feces or a hole in the wall.)

But remember, also, as we are still held in Earth's subtending conditions,

that acknowledgment of Mars' ground cannot end with a *certain amount of technical know-how* but must rather acknowledge that terrestrial knowledge *systems* are inadequate for an ongoing Martian world (see Olson 2018). This essay accepts settlement advocates' claim that current technology will allow humans to survive, initially, on or under Mars's surface. But to endure (even to thrive) demands knowing how to abandon thin, inadequate terrestrial epistem-ontologies in the face of the abundant, specific relations and rich, unique forms of relationality of this particular Martian world. It is to this end—the presumption of humans and nonhuman others seeking to thrive in a world that only looks apocalyptic or utopian from Earth—that I draw, orthogonally, from Nikki Giovanni in insisting that this Martian colony could be thought of as *A Good Thing.* This outlook is from the perspective of the creatures and things—however they got there, whatever their pasts, whoever they are, human or not—who must *know-how* to live by acknowledging Mars' ground if they plan on the next generation (of humans, microbes, plants, AI, concrete walls). And while—yes!—it is true that Mars never asked for these arrivants, from this speculative underground Martian world the question about Mars' own destiny, its rights, or its orbit may seem *hollow.* They are here, now, and from their perspective, Earth never asked for you, either. The question of what is *ethical* on Earth will not do under Mars because *ethics*—as Lakota star knowledge reminds us, from Earth—is always grounded in place, and in this place it flows into other, unlikely terrestrial domains so completely that its solo enunciation may feel odd on the Martian tongue.

And so,

the terrestrial distinction between *settler* and *Indigenous* may also be inadequate to Mars, equally needing to be jury-rigged in relation to Mars' ground; it may also not be the place to start. Lakota star knowledge would need to recalibrate to new star charts alongside the emplaced knowledge that terrestrial arrangements and constellations don't pertain from Mars. Giovanni's Black arrivants on Mars may have to forego the *fried chicken* she imagines as a first Martian meal, and *exploration* in a Maoli Kānaka Mars expedition would need to attune itself to wayfinding conditions that transform what *curiosity* draws attention to in this particular place. However, if not privileged on Mars, nonetheless the fundamental query among these decolonizing stances—about *where to start* an account of *humanness, nature, world,* or *universe*—may open to thriving imaginations of a vital, humanish place on Mars that demands a different starting place for both *colonization* and *decolonization* from this particular Martian point of view.

So, let's start again,

and speculate for a moment that among the always-tentatively settled world at Giovanni Prime, someone (or, more likely, someones) decide(s) to write a speculative, science fiction story. (Perhaps they will make a joke and call it *science f(r)iction.*)

But wait (you *still can*, they *can't ever*).

If this task seems frivolous from Earth, perhaps it would be valuable among them, an additional actant incorporated within their unfolding world to help them imagine new possibilities that work against the cargo of unacknowledged compatibility they carried here and that remains a burden. But *writing a story* (as much as an essay for a scholarly collection) would always already need to be acknowledged and incorporated as part of essential, non-deferrable, moment-to-moment world-continuing action upon which everything and everyone relies in this here-now.

So, if this/these yet-Mars-unborn author(s) reimagined Sax on Mars, they might start with the act of *crouching* and seek a whole, luxurious, and abundant Martian world in that stance. Perhaps they will start with domain-collapsing among the rudimentary, parsimonious, separated-out Terran things–actants–forces–states (*politics*, *muscle tone*, *chemistry*, *kinship*, *gravity*, *society*, *axial tilt*, *microbes*, *affect*, *stance*, *labor*, *atmosphere*) and make a sly joke about the provincial, suffocating human concerns that arose from Earth's jagged histories (*nature/culture*, *futurity*, *agency*, *humanness*, *destiny*, *justice*, *universality*, *exploration*, *intention*, *settlement*). But they cannot start with the outlandish one-thing-after-another of Terran *plots* that presupposes deferring conditions for human action or with the teleological presumption of *terraforming*. What might *stun* this Sax is not the terrestrial surprise of its animals on Mars but, rather, how observation itself is always implicated in the abundance of Mars' relations. What might *disconnect* him is recognizing the impossibility of *disconnection*, the spatiotemporal demands of *waiting* and of what and who is invoked in the phrase *trust me* in this here-now. *Above all*, what might *startle* this newly Martian Sax who must quickly return to the whole world, *below*, because radiation is that *human* is the least stable or interesting of these feeble, terrestrial obsessions.

But wait.

I cannot write this story. That would only return us to the common, containing, terrestrial loop of me-proposing and you-finding-the-gaps as—*meanwhile*—we debate or disagree, thrive or suffer, crouch in awe or hunch in pain in Earth's deferring conditions without immediate consequences for one another and everyone and everything else. I can only acknowledge where it might start from Mars, including—vitally—the breaking-off of its writing-now because some arrangement has emerged that cannot wait, that demands everyone's and everything's attention, so that continuing-to-write—*meanwhile*—could *end this specific world*, and not only produce speculations about worlding and world-ending in general. To write or know about a world, that is, is always simultaneously an ontological as much as an epistemological matter, a distinction that ceases to make sense under the red surface of Earth's closest, demi-habitable neighbor.

Acknowledging Mars' conditions as the starting place for a Martian science f(r)iction reveals that Robinson's account of Sax could only ever have taken place on Earth's surface, *after all*. It opens to an appropriately materialist acknowledgment of human engagements with the cosmos that reveals alien grounds as incompatible with multiple terrestrial materialities: the organic formation of *soil* in which one can simply *dig*; *crouching* as a neutral ability/stance from which to consider an unfolding world-universe; the universalisms of *geo-* in *geology*, *historical geography*, or *geophilosophy*; or the stability of *capitalism* or *star knowledge* or *socialism* or *middle passage* as if their unfolding were disconnected from the conditional arrangements of a single planet-world that subtends their truths, contradictions, and excesses. Such acknowledgments do not diminish vital questions about the consequences of human activity in space *on* or *for Earth* and its multitudes, human or nonhuman (Clark and Szerszynski 2021). But starting *from* Mars reveals—just beneath the surface of space settlement planning, critical outer space studies, and metropolitan science fictions—the material universe-fact that the most science fictional and the most utopian scenario of all is to presume the compatibility of the mess and beauty humans have made in Earth's homely conditions with the rest of the universe.

As for the residuum—as for what is in this essay for *us*, on Earth—there is merely this: An anthropology of a future, humanized outer space must begin by acknowledging the provincial grounds from which it starts. It must acknowledge that knowledge and experience of the universe (and the universal) formed on its ground is incompatible with starting to dig into what may become of humanness elsewhere in the cosmos. With this acknowledgment, *the universe* might have something to say back to Earth. But *meanwhile*, in writing about—as much as planning for—human presence beyond Earth, we are charged to start not only by thinking otherwise about Earth's futures but also from otherwheres.

5

Knowledge Frontiers

Shaping African Futures Through the Square Kilometre Array

Davide Chinigò and Cherryl Walker

Introduction

In this chapter, we address the role of science and technology in shaping a knowledge society in Africa through the case of the Square Kilometre Array (SKA) project in South Africa.[1] The SKA is a globally funded enterprise to construct the world's largest radio telescope; when completed, the infrastructure making up the array will be dispersed across multiple sites in Africa and in Western Australia. After a fiercely competitive bid process with Australia, South Africa won the right to host the major share of the array in 2012, at a site in the semi-arid Northern Cape Province, approximately 700 km northeast of Cape Town. Once completed (projected for some time in the 2030s), the SKA infrastructure in Africa will comprise thousands of interconnected antenna dishes, not only spread across the Northern Cape but also extending into eight other countries in southern, western, and eastern Africa. In examining this extraordinarily ambitious astronomy project as a terrestrial endeavor, we explore questions about the nature of the engagement between the global knowledge economy and "Africa," here seen as a system of representations mobilized through science and technology, as well as the social formations that emerge from such engagement. We contend that the lens of African studies can provide a critical perspective of the ways in which we unpack questions about the terrestrial implications of outer space research in social science.

There are many ways of looking at the role of science and technology in Africa. In an edited volume, Clapperton Mavhunga (2017) contends it necessitates a critically important prior question: What constitutes knowledge

Davide Chinigò and Cherryl Walker, *Knowledge Frontiers*. In: *Otherwhere Ethnography*. Edited by: Istvan Praet and Perig Pitrou, Oxford University Press. © Oxford University Press (2025).
DOI: 10.1093/9780197790885.003.0006

from Africa? For Mavhunga, this question not only involves reassessing the legacy of colonialism and the role of Western epistemologies in imposing particular notions of progress and modernity on the continent but also foregrounds perspectives on the continent's future that are shaped by Africans themselves.

Africa rising is currently a central trope framing the debate about science and technology among political elites in the continent. Following this view, science, technology, and innovation are centerpieces of modernity and key to repositioning the continent within the global knowledge economy. For example, the African Union's "Science, Technology, and Innovation Strategy for Africa 2024" (STISA) describes how "accelerat[ing] Africa's transition to an innovation-led, Knowledge-based Economy" is expected to fulfill important developmental challenges, including the "eradication of hunger and achieving food security; prevention and control of diseases; communication; protection of our space; live together—build the society; and wealth creation" (African Union Commission 2014, 10–11). Mavhunga (2017, 2), however, argues that strategic documents such as STISA 2024 are formulated around Western concepts and meanings, making science, technology, and innovation about the transfer of knowledge to Africa rather than the production of knowledge with Africa. He also highlights how the scientific revolution in Europe in the 16th and 17th centuries was profoundly entangled with Western imperialism and the colonization of the Global South. For this reason, he warns against thinking about knowledge systems, including Western and African epistemologies, as existing in isolation from each other.

Such concerns are central to debates about the meaning of decolonizing knowledge (Shipley 2010; Mbembe 2019). We agree that knowledge is co-produced and is the outcome of complex histories of domination, diffusion, and assimilation, in which power relations are always present. From this perspective, African societies are coming to science and technology not as outsiders but, rather, as participants in global processes of knowledge production and dissemination. At the same time, African societies are neither homogeneous nor egalitarian: South Africa itself is one of the world's most unequal societies, in large part because of its particular history of colonization and apartheid.

In this chapter, we center our discussion about the engagement of African societies with global science on the concept of knowledge frontiers. Knowledge frontiers are liminal spaces where scientific knowledge is negotiated in relation to other knowledge systems and contested visions of African futures as well as its pasts compete with each other. These are spaces in which new territorialities refashion power and authority and new patterns of accumulation reshape dynamics of inclusion and exclusion, at different scales.

Astronomy in South Africa is a prime example of the tension between the universalizing claims of science and technology and their unequal impacts across different scales. As noted by historian Saul Dubow (2019, 2),

> Astronomy's principal object of fascination is the universal, non-human domain of the cosmos. Yet the pursuit of astronomy was from the outset firmly conditioned by the need to prove its practical utility: navigation, meteorology, time-keeping, tidal observations, hydrography and boundary mapping. . . . Another theme is the interplay between overseas and local expectations and demands or, in other words, the balance between northern and southern hemispheric power.

Within astronomy, the SKA is a particularly revealing example of knowledge frontiers as sites in which African engagement with the global knowledge economy takes place. The complex and futuristic infrastructure of the SKA encompasses a wide range of elements—what we have elsewhere referred to as an "assemblage" (Walker and Chinigò 2018)—that extends from the cosmic dimension of its scientific agenda through national capitals in the north and south to the small, marginalized towns in South Africa's Northern Cape Province adjacent to the SKA's core site.

In this chapter, we address three overlapping aspects of the knowledge frontiers implicated in this major global science project. The first concerns how "Africa" emerges as an epistemic object through global science and technology. The second is about the significance of the SKA in shaping not only North–South relationships but also the global and continental leadership ambitions of post-apartheid South Africa, at a time when concerns over decolonizing knowledge are challenging postcolonial relations. The third is about the impact of the SKA in the area of South Africa where its core infrastructure is being built, as a major land use change entailing the creation of a large astronomy reserve in a politically and economically marginalized region of the country.

We conclude that this case study shows that African engagements with the global knowledge economy and the representations of "Africa" within that do not reflect a linear trajectory of change. The notion of knowledge frontiers allows one to probe important dynamics of inclusion and exclusion that raise challenging questions about the role of outer space science and technology in postcolonial Africa. While science and technology enable selective dynamics of inclusion, the opening of new knowledge frontiers also involves the establishment of new boundaries of exclusion and inequality, through new configurations of authority and power.

The chapter draws on a number of research projects conducted since 2016 within the SARChI Chair in the Sociology of Land, Environment, and Sustainable Development at Stellenbosch University in South Africa.[2] In the next section, we briefly review how we are deploying the concept of knowledge frontiers in this chapter. Thereafter, we discuss the three aspects we wish to highlight in turn.

Knowledge Frontiers in Africa

The concept of the frontier was first systematized in 1893 by the American historian Fredrick Jackson Turner to indicate a geographical space of continuous transformation shaped by efforts of the center (the "empire") to expand a polity's domination and territorial control (Turner 1921). For Turner, the western frontier of the United States was a space characterized by the encounter between "savagery" and (western) "civilization"; its transformation was shaped by the agents of this mission to civilize the savages, control the natural environment, and secure access to resources. In Africanist literature, the notion of frontier is often associated with Igor Kopytoff's *The African Frontier* (1987), in which the frontier is defined in terms of the multiple processes of domination, pacification, and enculturation of precolonial African societies as they were inscribed in the colonial project. In South Africa, Martin Legassick's very influential 1972 essay, "The Frontier Tradition in South African Historiography," represented a critical departure from the then dominant liberal tradition in South African history; here, the frontier was characterized in terms of the broader system of racial segregation embedded in first the Dutch and then the British colonial projects and their underlying class dynamics.

Several historians have emphasized that while characterized by conflict, competition, and contestation, the frontier is also a zone of creative cultural production and experimentation (Donnan and Wilson 1999). In South African historiography, debates over the role of the Cape Colony's eastern frontier in producing an Afrikaner identity highlight the myth-making value of the idea of the frontier for national elites. Frontiers are fluid sites through which certain views of the past can be solidified and institutionalized by the political center. In his seminal study of the northward's expansion of the boundary of the Cape Colony in southern Africa in the 17th and 18th centuries, Nigel Penn (2005) reconstructed the history of this "forgotten" frontier and the often violent interactions between White *trekboer* settlers and Indigenous Khoisan people this entailed. Penn characterizes this frontier as both a symbolic and a physical site where older identities were crushed but new identities forged.

More recently, Korf and Raeymaekers (2013, 10) have noted that "frontiers are not necessarily boundaries, but rather political spaces with distinct spatialities of rule and sovereign power." Frontiers are thus dynamic sites where competition for the institutionalization of power and authority finds its expression through territorial processes. Following Peluso and Lund (2011, 668), the frontier signifies processes of change from one regime of governance to another; frontiers are "sites where authorities, sovereignties, and hegemonies of the recent past have been or are currently being challenged by new enclosures, territorializations, and property regimes."

The broader field of political ecology embraces an understanding of frontiers as spaces forged at the intersection between social and ecological change (Cavanagh 2018; Nygren and Rikoon 2008). Analyzing the dynamics of sedentarization in the Ethiopian pastoral frontier, Korf et al. (2015, 882) discuss the frontier as a process of land appropriation that is "co-produced by the projection of state sovereignty into the drylands and by the increasing commodification of . . . land-based pastoralist resources, the drivers of which originate both from within and from outside pastoralist societies." This reflects an understanding of the frontier as a space in which localized knowledge systems play an important role in the co-production of authority from below, generating complex spatialities of power and outcomes not intended by the planned interventions from above.

Although most of the literature touches upon the temporal dimensions of the frontier, time is seldom discussed in a systematic way. A partial exception is the work of Steffen Jensen and Olaf Zenker (2015), who, in discussing the South African homelands as "the loose end of apartheid," employ the notion of frontier to characterize intense zones of contestation where the future rather than the past of post-apartheid South Africa will in part be negotiated. From this perspective, frontiers can be regarded as spaces where multiple temporal trajectories are negotiated and contested.

Building on this body of literature derived from African studies, in this chapter we employ the notion of knowledge frontiers to describe the complex and nonlinear dynamics through which global science and technology are territorialized in Africa. Knowledge frontiers are liminal spaces in which science and technology not only shape visions about a future for humanity as one but also are our entry point to discuss inequality, competition, and conflict across different scales, as well as the cultural production of how Africa is constructed as an epistemic object. We contend that reflecting on the notion of frontier from the perspective of African studies literature can provide new insights on the terrestrial implications of outer space research in the social sciences, as well as troubling claims that the concept of frontier is of little analytical help when applied to conceptualization of outer space.[3]

The SKA and the Frontiers of Global Scientific Knowledge

As already indicated, the SKA is a globally networked scientific initiative that brings together national governments and consortiums of research institutions from numerous countries throughout the world. Its scientific scope encompasses questions at the forefront of global scientific knowledge, which its advocates use to represent the project as operating on the frontiers of progress for humanity at large. The ambitious scope of the project is captured in the project's mission statement (SKA 2016, 4):

> [The SKA will] tackle some of the fundamental scientific questions of our time, ranging from the birth of the Universe to the origins of life. How did the Universe evolve to its current form? What is the mysterious nature of Dark Energy? How and when did magnetic fields arise to influence the Universe as they now do? Was Einstein right about gravity? Can we understand the nature of the gravitational waves on which the Earth surfs? How does one make a planet from pebbles in space? How prevalent are the molecular building blocks of life across the cosmos? Are we alone in the Universe?

Answering these questions is expected to revolutionize not only the field of astrophysics but also the natural and physical sciences and the human sciences more broadly. These claims are captured in the notion of the SKA as embodying an agenda of "transformational science," meaning science with major implications beyond radio astronomy and engineering. Following the SKA Director of Science,

> The SKA will revolutionize our understanding of the Universe in many ways, . . . and will also aim to answer whether Life exists elsewhere in the Universe. Should it answer that question, it would have immense repercussions from science to religion and philosophy. We should also be ready for more unexpected things the SKA could discover. The SKA's ability to fundamentally influence many fields is what makes it a transformational science machine.[4]

However, as noted by Messeri (2016), scientific innovation takes on different meanings and values in different contexts. How is a project aimed at exploring a boundless universe also entangled in representations of the human, more particularly for the purpose of this chapter, of Africa, the continent in which much of the infrastructure required to realize this vision is located? How are representations of the universe territorialized in the frontier field of astrophysics, which in this case means how do the spokespeople for SKA depict the SKA in "Africa"?

There is much that can be said on this subject, but for the purposes of this chapter, we focus on the depiction by those driving the project in South Africa that is presented in Figure 5.1. This billboard, erected after South Africa had won its hosting rights on the outskirts of Carnarvon, one of the small towns closest to the SKA's core site, is a representation of what the infrastructure will look like once completed.[5] It depicts the semi-arid Karoo landscape of the Northern Cape covered with countless satellite dishes.[6] In the starry night sky above the iconic flat-topped hills of the Karoo, it is possible to distinguish the Milky Way, indicating the connection of the landscape—"Africa" according to the title on the billboard—with the center of our galaxy. In the foreground, a large, more detailed representation of a satellite dish pointing at the night sky signals the project's focus on exploring the universe while even closer a meerkat, a small, endearingly photogenic rodent found locally and throughout southern Africa, surveys the scene knowingly.

On the billboard, this animal serves a dual role. For those in the know, it invokes South Africa's 64-dish MeerKAT radio project,[7] the precursor to the international SKA proper that South Africa developed once it had won the 2012 bid. This array, inaugurated in 2018, was itself built on the back of a smaller 7-dish telescope, the Karoo Array Telescope or KAT-7, which was developed in the early 2000s in support of South Africa's bid for the SKA. The name MeerKAT is a clever play on the Afrikaans word *meer*, meaning

Figure 5.1 SKA billboard in Carnarvon, 2016.
Source: Davide Chinigò.

more—that is, more KAT dishes to take the project forward. The MeerKAT array will be integrated into the broader SKA international project, which in South Africa involves the construction of approximately 133 more antennae dishes under the auspices of the international Square Kilometre Array Organization (SKAO), the first of which are currently under construction.[8]

While reflecting important features of the SKA's understanding of itself as on the frontiers of global knowledge, this billboard also raises intriguing questions about what is not present. Most strikingly, where is the human? How does human history, the history of this region of South Africa, relate to the history of our universe? As we have argued elsewhere, in choosing its South African site, SKAO emphasized the locality's emptiness: the presumed absence of history, despite this area being home to |Xam hunter-gather society for millennia (Walker 2019; Parkington et al. 2019). The billboard offers a vision of how a future driven by science and technology will transform the empty landscape surrounding the core site of the telescope—will transform "Africa." In speaking to ideas of human progress through outer space exploration, is it an accident that no people are included in the picture?

This is where the dual role of the meerkat in the billboard comes to the fore. While suggestive of the domination of the project over nature, the meerkat surveying the scene provides a heavily anthropomorphic subject but with no specific identity—a subject without race, class, or gender in a place where all three are crucial determinants of power. In this way, the human subject has been carefully redefined in relation to the "SKA in Africa," absented from the actual site where the infrastructure is being constructed but projected into the future that the project will enable. At the same time, the contested history of the actual place where the array is being built is erased, as are the contestations around the contemporary land use changes involved in converting this area from commercial sheep farming to an astronomy reserve (Chinigò 2019; Parkington et al. 2019; Terblanche 2024).

On the Frontier of North–South Relationships

This section elaborates on the SKA knowledge frontier as a space where international relations are negotiated, both those captured under the umbrella of north–south relationships and those involving the African countries earmarked for hosting components of the array. This points to the role of science and technology in shaping representations of Africa in terms of the national development ambitions of African countries (Donnan and Wilson 1999; Korf and Raeymaekers 2013).

While the core of the SKA infrastructure is currently under construction in two sites in the southern hemisphere—Australia and South Africa—the international consortium involved in the project is based firmly in the North. As of July 2023 (Figure 5.2), 16 countries were involved in the SKA at the government level or represented as observers—Australia, Canada, China, France, Germany, India, Italy, Japan, the Netherlands, Portugal, South Africa, South Korea, Spain, Sweden, Switzerland, and the United Kingdom—with the potential of other countries joining over time.[9]

The SKA project was initially conceived in the early 1990s, with the first international Large Telescope Working Group set up in 1993. Although ownership over the idea behind the scientific design of the SKA is disputed,[10] in 2000 eleven countries signed a memorandum of understanding to establish the International Square Kilometre Array Steering Committee. From

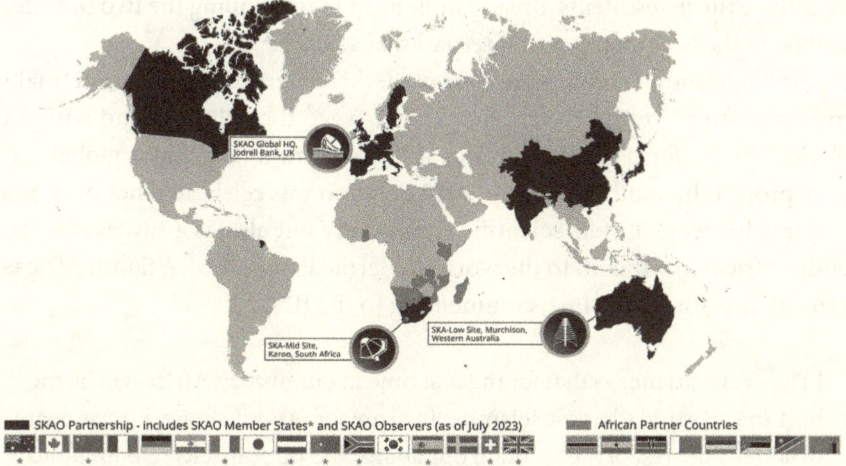

Figure 5.2 The SKA international consortium.
Source: SKAO.

December 2011 to March 2019, the project was led by the SKAO, a nonprofit organization registered in the United Kingdom, with the international headquarters located at Jodrell Bank Observatory outside Manchester. On March 12, 2019, a new phase in the governance of the project was initiated when the Square Kilometre Array Observatory (SKAO) consortium was reconstituted as an intergovernmental treaty organization at a signing ceremony involving seven member countries in Rome (with several other countries expected to sign the treaty in the future). In this form, the SKAO has a special international status, similar to that of the European Organization for Nuclear Research (CERN). Figure 5.2, which shows the SKAO as of February 2024, reveals important dimensions of the North–South relationships embedded in this international body.

As already noted, the initial bid process was characterized by tense competition between South Africa and Australia, with many experts assuming that the long tradition of radio astronomy in Australia and the country's international reputation as a stable modern state would inevitably carry the day. However, in May 2012, the adjudicating panel of experts announced the dual-site outcome in which South Africa was awarded two-thirds of the project and Australia one-third. To the surprise of many, the South African bid was judged technically superior in many respects, but the adjudicating panel motivated its decision to split the award on the grounds that this would maximize the investments already mobilized in developing the two bid sites and be in the best interests of international science.[11]

Although some critics have disputed this interpretation as not making sense from a purely scientific perspective,[12] the split decision still put "Africa" in the form of South Africa at the forefront of a major global science project. In South Africa, the 2012 decision was celebrated not only as a major achievement of its scientific community but also as a success for the entire African continent. In the words of Bernie Fanaroff, SKA South Africa's project director at the time, commenting to the BBC,

> [The decision] means that for the first time in our history, Africa will be the host to the world's largest scientific instrument. And it shows a great deal of faith by the rest of the world in our ability and our capacity to both build and operate such a sophisticated instrument. It also reflects the recognition in Africa of how important science and technology is to our future.[13]

This claim is remarkable for its representation of North–South relationships: For the first time in history, a global project in fundamental science will be undertaken in Africa, enabling a historic reversal of colonial relations. At the same time, South Africa's international success is a central theme in the national legitimation of the country's bid to host the project: A central dimension of this narrative is that science has an intrinsic developmental ethos, with the bid award marking a moment of rupture with representations of Africa as the continent lagging behind the rest of the world.

However, the development of astronomy in South Africa is deeply implicated in the history of colonization across the continent. The roots of astronomy in the country can be traced back to the establishment of the Royal Observatory by the British government on the outskirts of its colonial outpost of Cape Town in 1820. Historian Saul Dubow (2019) has noted that astronomy was the first scientific discipline to root itself institutionally in South Africa, with observatories not simply prized viewing platforms for European astronomers but also major sites for the production of information that was central to the imperial project, such as navigation, map-making, and time-keeping. In the colonial imagination, astronomy also embodied the civilizational force of modern science, the latter a significant marker of the boundary between the civilized and the noncivilized that legitimated the colonial project in Africa. Subsequently, astronomy in South Africa came to be nationalized in the service of the apartheid regime from the mid-20th century, although international collaborations with astronomers in the United Kingdom and the United States continued under the surface during the period when South Africa was isolated internationally after 1977 (Dubow 2019).

After South Africa's transition to formal democracy in 1994, the colonial roots of astronomy were downplayed as this field of scientific knowledge was embraced as a force for transformation by the ruling African National Congress (ANC) in the late 1990s and early 2000s. The successful promotion of the SKA project in the country at this time was bound up not only with ideas about internal transformation but also with the vision of an African Renaissance animating the Mbeki presidency (1999–2008). Following its advocates, the African Renaissance was based on

> the need to empower African peoples to deliver themselves from the legacy of colonialism and to situate themselves on the global stage as equal contributors to . . . human civilization. Just as the continent was once the cradle of humanity . . . this renaissance should empower it to help the world rediscover the oneness of the human race. (Mavimbeka 1998, 31)

Similar principles pointing specifically to the role of science and technology in fulfilling social and development objectives within South Africa can be found in South Africa's 1996 "White Paper on Science & Technology" (Republic of South Africa [ROSA] 1996, 16):

> Scientific endeavour is not purely utilitarian in its objectives and has important associated cultural and social values. Not to offer them would be to take a negative view of our future—the view that we are a second class nation, chained forever to the treadmill of feeding and clothing ourselves.

The SKA was positioned as the epitome of this narrative. Through the SKA, "Africa" would be enabled to play a leading role in the production of scientific knowledge, with numerous positive benefits cascading from this investment for society in general and the national economy in particular.

From this perspective, the map of the SKAO (see Figure 5.2) shows South Africa taking its rightful place as an equal partner with its international peers in the project. However, another way of interpreting the map is that it illustrates the continuation of the dynamics of domination emanating from the history of Europe's colonization of Africa. The global headquarters of the project are based in the United Kingdom, the colonial power par excellence, and European countries are providing the bulk of the considerable financial resources required to build the telescope. Furthermore, considering that the primary output of the infrastructure in Africa will be digital data that will be transmitted to the North and accessed by countries proportionate to their financial contributions to the project, the map reflects the continued reproduction of unequal North–South relations, not their reorganization.

Yet the picture is still more complicated than this. The absence of the United States and Russia and the presence of the emerging global powers of China and India index a refashioning of postcolonial dynamics within new geographies of global power—one in which South Africa's own status is uncertain. Furthermore, a closer look at the African dimensions of the project provides a further perspective on the contours of North–South relationships entangled in the SKA. As already noted, while the core site of the African infrastructure is currently being built in South Africa's Northern Cape, plans for Phase Two of the project call for this core site to be linked to outstations in Botswana, Ghana, Kenya, Madagascar, Mauritius, Mozambique, Namibia, and Zambia. However, South Africa is the only full member country in the international SKA consortium; the other African countries in the project qualify as "partner countries," meaning that they will host part of the infrastructure but will not commit funds to the project nor be directly involved in its scientific production.

Although some scientists we have interviewed have expressed doubts about the feasibility of Phase Two ever being fully implemented—because of both the technical and the sociopolitical difficulties involved—currently the involvement of the partner countries is taking place under the umbrella of a program called the African Very Large Baseline Interferometer Network (AVN) that is hosted by South Africa's national office for radio astronomy, the South African Radio Astronomy Observatory (SARAO). The AVN, funded largely through the South African Department of International Relations and Cooperation, is aimed at developing the skills and institutional capacity needed in the SKA's African partner countries to participate in the broader project. In addition to study opportunities for qualified students, AVN projects have included the development of a training facility in Ghana, based on the conversion of a telecom antenna to radio telescope in the Accra suburb of Kuntunse (completed in August 2017), with a second training facility currently planned in Madagascar.

When inaugurated in 2018, the Ghana telescope was celebrated as another success story for Africa. More recently, problems with costs of operation and maintenance of the facility have cast a shadow on the extent of such success story (Merron and Lamoureaux 2024). Nonetheless, the continental configuration of the SKA infrastructure, with the partner countries relegated largely to hosting further infrastructural sites, points to the ambiguities around the role of South Africa within Africa. Although the AVN program is constructed around a narrative of the future of the continent inspired by the values of pan-African humanism that underpin the African Renaissance, it also reproduces understandings of South Africa as the continent's "exceptional giant," the African member of the BRICS countries, and (aspirant) sub-imperial power.

The Karoo Resource Frontier

In this section, we highlight the link between the SKA as a knowledge frontier and its view of its core site in the Northern Cape as a resource frontier (Walker and Hoffman 2024, 2). This brings into sharp focus the significance of the project as a significant terrestrial land use, requiring a major land-use change (from commercial sheep farming to astronomy) in South Africa's Karoo region. From this perspective, the knowledge frontier represented by the SKA is also a project of land appropriation entailing various processes of rescaling, enclosure, and land alienation, framed by expectations of socioeconomic development (Korf and Raeymaekers 2013, 30).

As background to this discussion, it is necessary to understand that radio astronomy is an uncompromising land use in terms of the operating environment it requires. This is because it is based on the detection of very faint cosmic radio signals from the universe that are drowned out by the radio emissions produced by numerous technologies now taken for granted as part of daily life, including mobile phones, microwaves, Wi-Fi systems, and gas-driven vehicles. Because all of these technologies threaten the optimal performance of the SKA telescope, in 2007 South Africa passed legislation, the Astronomy Geographic Advantage Act (ROSA 2007), that granted extensive powers to the Minister of Science and Technology to restrict activities that would produce unwanted radio emissions in all of the Northern Cape Province outside its capital city of Kimberley. This legislation provided the framework within which a number of nested "astronomy advantage areas" (AAAs) were subsequently proclaimed around the SKA's core site where MeerKAT was being constructed. The three Karoo Central Astronomy Advantage Areas span an area of 100,000 km^2—approximately 10% of South Africa—in which restrictions on any activity deemed to threaten the operations of the SKA become increasingly tighter the closer one gets to its core site (Figure 5.3). This has become a major source of opposition and resentment among local residents. A major concern was that alongside the most advanced data collection and communication technology in South Africa, the use of communication technologies that local people have come to rely on, such as mobile phones, would be constrained.

A further controversial initiative that took place alongside the proclamation of these AAAs was a compulsory program of land acquisition by the National Research Foundation of South Africa, on behalf of the SKA South Africa, between 2016 and 2019.[14] The program entailed the purchasing of 40 farm parcels for the SKA's core site, in addition to the two farms initially bought in 2008, which many local farmers had assumed would constitute the total land required. Encompassing a total area of 135,245

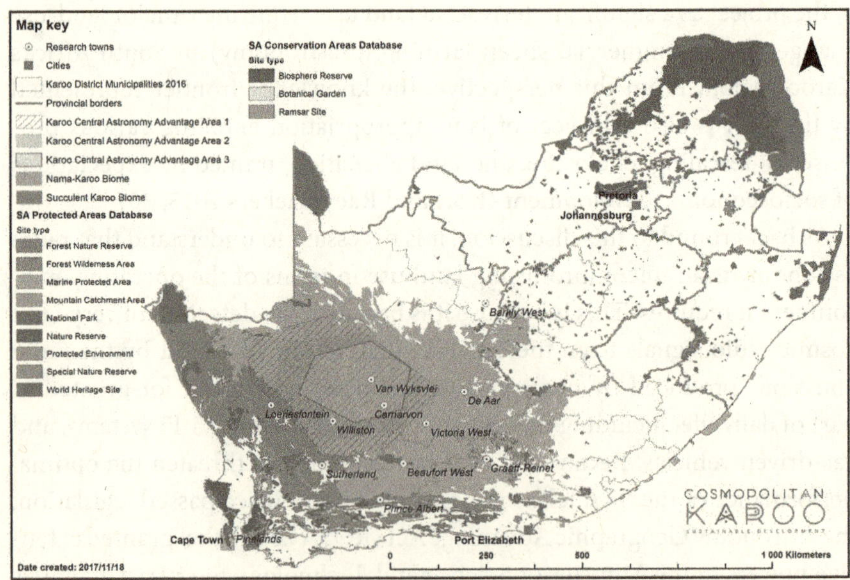

Figure 5.3 The Karoo Central Astronomy Advantage Areas.

Source: SARChI Chair in the Sociology of Land Environment and Sustainable Development, Stellenbosch University.

ha, in March 2020 this SKA core area was officially proclaimed a national park dedicated to scientific research—the Meerkat National Park (ROSA 2020). The land acquisition program also generated significant tensions in the local towns and farmlands surrounding the SKA core site. A major concern voiced most strongly by local farmers, all of them White, was the impact of the new astronomy reserve on their livelihoods as well as on the local economy more broadly, which since the late 19th century has been based on commercial sheep farming (see Walker and Chinigò 2018; Terblanche 2024).

On November 1, 2017, SARAO and the South African Environmental Observation Network (SAEON) signed a memorandum of agreement to define the integrated environmental management plan for the SKA project, envisaged in this agreement as a nature reserve serving "as a science park for both astronomy and ecology."[15] Commenting on the agreement, Johan Pauw, Managing Director of SAEON, explained, "South Africans will experience significant value-addition from this extraordinary synergistic relationship between two seemingly opposing sciences. For the first time ever, Radio Astronomy and Earth Observation research infrastructures will be collaborating in South Africa to ensure sound environmental management."[16] A follow-up of the agreement with South Africa's National Parks Authority, SANParks, provided for the development of the SKA core area as a national park—Meerkat National Park—in which radio astronomy would be the premier land use.

Although the full story of the astronomy reserve is yet to be written, its declaration highlights a further aspect of the SKA's knowledge frontier—in this case, its alignment with the conversion of the land on which the array is being built into a new resource frontier. What is remarkable about the land use change involved in this new Karoo resource frontier is its legitimation by means of a mixed discursive repertoire involving environmental conservation and the pursuit of science, hence reflecting new dynamics around the territorialization of central state power in this historically marginal area of South Africa. These new patterns of authority and power intersect in complex ways with the history of the region. They raise fundamental questions about the what and the who that science and technology include and exclude, and the power relations involved in making these decisions.

Because of its aridity, the Northern Cape is the largest but most sparsely populated province of South Africa, covering almost one-third of the country landmass but accounting for only 2% of the population. Most of its residents are concentrated in small towns widely separated from each other and characterized by significant social challenges, including widespread unemployment and dependence on government social grants (Walker and Vorster 2024). By the time the "forgotten frontier" described by Nigel Penn (2005) had closed in the second half of the 19th century, the colonial history of racialized land dispossession and appropriation in this region had laid the groundwork for the consolidation of White-owned commercial agriculture, in which the surviving descendants of the Indigenous Khoisan people were either displaced or turned into a subjugated class of laborers. Subsequently, this racialized social hierarchy crystallized into a minority of White and a majority of "coloured" people, the latter term the official designation for virtually all the people deemed "not white" under the apartheid regime from the 1950s.

The historical marginality of the Northern Cape vis-à-vis South Africa's political and economic mainstream continues today. A central element of continuity reflected in the SKA's reconfiguration of this historical frontier zone as a new resource frontier is the tension between local residents and metropolitan interests. This can be seen in the SKA's characterization of this region as an empty space that is ripe for development in the name of both national and global interests. The astronomy reserve, discursively constructed at the intersection of the national pursuits of astronomical science and environmental conservation, reproduces the historical construction of the Cape Colony's northern frontier as an empty space for colonization and development on behalf of metropolitan elites. The projection of emptiness from the political center recalls the understanding of resource frontiers as spaces that await civilization, empty of history and local territorial claims (Donnan and Wilson 1999).

At the same time, as noted in the previous discussion, through the SKA this new resource frontier is also projected as a space of desire and imagination (Korf and Raeymaekers 2013), full of promise for a better future in terms of not only its development potential but also its reconfiguration of the place of "Africa" in the world.

Conclusion

In this chapter, we have identified some of the heterogeneous and contradictory dynamics underpinning the role of science and technology in shaping a knowledge society in contemporary Africa, through a case study of the SKA project. We have deployed the concept of knowledge frontiers to elucidate the selective and fragmented ways through which knowledge is produced discursively and territorialized at different scales.

We have argued that the SKA illuminates three important dimensions of what we have described as knowledge frontiers. The first is the territorialization of global knowledge in science and technology as it is inscribed in representations of Africa from within and without the continent. Meanings and values attached to the pursuit of science—channeled through Western epistemologies and their universalizing claims but claimed as co-produced by Africanist scholars—raise important questions about what constitutes knowledge from Africa, beyond the transfer of knowledge to Africa. The second dimension concerns the nature of North–South relationships that are inscribed in the SKA project as a complex international consortium of governments and research institutes throughout the world. Although it is claimed that the pursuit of scientific knowledge will liberate the continent from the legacy of colonialism, the project also reflects continuities with the history of northern domination over the south that is inscribed in the colonial project. At the same time, the role of South Africa within this project brings to the fore the tensions between the pan-African humanism informing the idea of the African Renaissance and intracontinental competition for power and hegemony. Here, the idea of knowledge frontier serves as a complex metaphor for sites of struggle over representations of not only the continent's past and future but also desired development paths.

The third dimension invokes more conventional notions of the frontier as a space in which new territorialities are inscribed on actual social and physical places. The establishment of an astronomy reserve to serve as the core infrastructural site of the SKA in South Africa's Karoo has involved a significant land use change that is tied to the redefinition of public authority in this part of contemporary South Africa—that is, one involving an historical shift away from White commercial farming to the pursuit of science by the post-apartheid ANC government. In exploiting this new resource frontier, the Karoo has been re-presented as an empty space ripe for development in the service of national and global priorities.

These three dimensions show that African engagements with the global knowledge economy do not involve a linear trajectory of change. The knowledge frontiers opened up by the SKA are a prominent example of how Africa as an epistemic object is currently redefined via science and technology, and outer space research, at different scales. These representations coalesce and contradict each other without any one of them fully extinguishing the others. The territorialization of these representations reflects the redefinition of authority and power and new dynamics of inclusion and exclusion. Although engagement with the global knowledge society has opened up new opportunities for continental elites, it is also producing complex transformations of people's lived experience across different scales on the ground.

PART II

OTHERWHERE ETHNOGRAPHIES

Analogs, Instruments, Artifacts, Viruses, and Vegetables

6

Between a Rock and a Hard Space

On Instrument Time and Communication in BASALT

Zara Mirmalek

Introduction

The National Aeronautics and Space Administration (NASA)'s Apollo missions (1969–1972) produced six cases of humans working in situ on the Moon. These NASA personnel, cumulatively 12 male astronauts (Ackman 2003; Chaikin 1998; Nolen 2002; Weitekamp 2004), gathered data including images, geologic samples, and physical experiences (body and technology) of interacting with the Moon's environmental features and of communicating with other astronauts and with humans on Earth. In the last three Apollo missions, a lunar roving vehicle was added to the team (Swift 2021). It was one of the largest technologies that astronauts were able to practice using in advance while still on Earth at sites that were deemed similar in geologic composition to the Moon.

If it is an understatement to say that human corporeal in situ access to extraterrestrial worksites is limited, then what can be said about access to these sites for ethnographic engagements? Indeed, texts within this volume and elsewhere (e.g., Clancey 2002; López 2011; Marcheselli 2022; Messeri 2014, 2016; Olson 2018; Ormrod and Dickens 2016; Praet and Salazar 2017; Szolucha et al. 2022; Timko et al. 2022; Traweek 2010) speak to this question. In this chapter, I draw from ethnographic data collected as an observer-turned-participant-observer within a multiyear program, funded by NASA's Science Mission Directorate, Planetary Science & Technology Through Analog Research, called Biologic Analog Science Associated with Lava Terrains. Its acronym, BASALT, signifies connection vis-à-vis a geologic object found on both Earth and Mars: Basalt is the name of a volcanic rock that forms from cooling lava. BASALT's objectives included envisioning and testing a work system for human–robotic exploration on Mars.

Zara Mirmalek, *Between a Rock and a Hard Space*. In: *Otherwhere Ethnography*. Edited by: Istvan Praet and Perig Pitrou, Oxford University Press. © Oxford University Press (2025). DOI: 10.1093/9780197790885.003.0007

The use of analog sites for process and tool development for science and exploration on the Moon, planets (Mars), and asteroids has come to be known simply as "an analog." Analogs do not promise exactitude; it is acknowledged that there are differences between an extraterrestrial site's conditions, such as Mars, and a terrestrial site's offerings of Mars-like conditions. NASA space science researchers conducting analogs are themselves responsible for articulating what conditions constitute these gaps and if and how they will address conditions (e.g., Keeton et al. 2011). However, the valuing or tabooing of talking about these differences should be kept in sight.

Built on an Apollo tradition of employing analog sites to train astronauts, analogs do not require participants to have an impending trip to outer space (or low-Earth orbit). In this chapter, I draw from ethnographic data collected with the BASALT program, which included three expeditions (in a 2-year period, 2016 and 2017) to volcanic terrains serving as analog sites for Mars. BASALT's goals were to yield new scientific understandings about the geology and biology within volcanic terrains at two Mars analogs sites in the United States and to provide recommendations for human–robotic science exploration on Mars (Beaton et al. 2017; Johnson 2016; Kaplan 2016; Lim et al. 2019). Sixty collaborators had varying analog experiences, and all were engaged in research that held a relationship to planetary science. Categorically speaking, BASALT's team was multidisciplinary organized under three "pillars": science (e.g., biologists, geologists, and volcanologists), technology (e.g., data management software and telecommunication relays), and operations (e.g., digital systems for information and knowledge relay, and performance measurements).

My role as an ethnographic observer on BASALT began with an unexpected exchange with BASALT's Principal Investigator (PI) Darlene Lim at a biannual meeting of the National Oceanic and Atmospheric Administration's Ocean Exploration Advisory Board in Narragansett, Rhode Island. As a board member, Dr. Lim was seated at the conference table that took up most of the large room. I had driven down from Boston, Massachusetts, for a seat in the public audience on the chance that I might be needed to further describe some ethnographic research (Bell et al. 2015) that was featured in the presentation of another board member. When introduced and asked to provide a definition of ethnography and its use in a study on human–robot teams using telepresence to conduct science in the deep ocean, I did so and added comments on qualitative methods and understanding science teams as multidisciplinary, multicultural. At the meeting break, Dr. Lim expressed to me a shared understanding that it mattered to understand a science team's multicultural composition. She described some of her analog research (e.g., Lim et al. 2011), and we continued the introduction a few weeks later over the phone. She invited me to BASALT's first of three expeditions, in Idaho, for which there were some open spaces, given that all BASALT members were unable to participate. Always ready to engage in research from which to learn more on the construction and cultures of the scientific knowledge production, I accepted. The next occasion that I would see her in person was in a Mars analog field.

While reading BASALT's funding proposal, the first thing that caught my attention were the familiar names of some of the co-investigators with whom I had worked on the NASA Mars Exploration Rovers (MER) mission in 2003 and 2004, including Athena Science team PI Steve Squyres (Squyres 2005; Squyres et al. 2003). On the MER mission, robots were operated on Mars to collect and send images and instrument data for scientists to review on Earth and respond with next-step instructions on a daily schedule. The temporal framework for this distributed work was arranged to give the MER science team approximately 10 hours to complete their work (involving multiple science and engineering teams, data interpretation, and team decision-making) before handing over data collection requests to MER engineers, who then checked and translated them to send as commands to the rovers before the workday's end. It was an arrangement that operated with a set of time–work relationships encompassing temporalities across planets, institutions, and history (Mirmalek 2020).

Before BASALT, my knowledge on analogs was slight. In the months prior to MER's nominal mission (January–March 2004), conducted at NASA/Caltech Jet Propulsion Laboratory in Pasadena, California, I ran into two fellow NASA Ames researchers, Maarten Sierhaus and Chin Seah, also traveling between NASA centers. Seated at the Burbank airport, Dr. Sierhaus talked about a project he was working on (Sierhaus et al. 2000) and a Mars analog in Utah, showing me a video of two people suited in white coveralls carrying black rectangular packs trekking through reddish-brown terrain. Later, during MER, I heard a snippet about a Mars analog in the Haughton Crater on Devon Island in the Arctic from another ethnographer, Bill Clancey (Clancey 2002, 2012; Lee 2002).

The composition of the BASALT community posed an interesting question on reproduction and work practice among communities considered harmonious but also incongruently distinct given situated differences on technology norms (e.g., digital media), demographics, institutional goals, and public accountability. How might the work practices from an actual mission with robots in situ at the field site be carried through its participants to shape creating work practices in an analog for a future mission with humans and robots in situ at the field site? Although I could not add a research question into the project goals, I proceeded to take the opportunity to be present and conduct ethnographic fieldwork as an observer.

My first foray with BASALT was followed by an invitation to the second expedition (fall 2016) and then the third (fall 2017) and a growing cultural understanding of a research field that is simply called an "analog" at NASA. Primarily designated as an observer during the Idaho expedition, I lent a hand when I could and also produced some analysis that I shared in conversation and some that I published (Mirmalek 2017). A couple of days into the second expedition, I observed some interactions that would lead to shifting into participant observation. In this chapter, I go through some of the context and grounding that supported my participation which interrupted BASALTers shortening the trial use of an instrument.

BASALTers' Rock Places

The BASALT team, "BASALTers" as they called themselves, used two sites in the United States as Mars analog sites: the Craters of the Moon National Monument and Preserve in Idaho (COTM) and Hawai'i Volcanoes National Park (HVNP). They utilized COTM for the first of three expeditions in June and July 2016 and HVNP for the second in November 2016 and third in November 2017. Calendar days for each field expedition were counted at 21 days; yet this is an inadequate accounting because it leaves out the prefieldwork that requires scheduled and unscheduled time for preparing for events in the field, for dozens of people, provisions, and tools. (On the use of national parks for space analogs, see Cockell 2007.)

BASALT referred to the project's time in the field as "deployments" and occasionally as "campaigns" or "expeditions." Having little feel for military terms and more experience with expeditions undertaken for curiosity, science, and geopolitics, I continue to use "expedition" to describe a dynamic event involving a temporarily organized group of people working to acquire (e.g., goals, material, and claims) in an outside environment through the use of tools and processes.

BASALT was a multidisciplinary and international team of approximately 60–80 (counting guests) people with varying professional backgrounds (e.g., volcanology, geobiology, geochemistry, geophysics, astronomy, aerospace engineering, biomechanics, robotics, computer science, and planetary science) and varying years of professional experience (e.g., student, early career, and senior; ages 17 to 60-plus years). The total number varied slightly between expeditions, increasing with additional overall participants but maintaining the primary science team. Their institutional affiliations included government and university research institutions (e.g., Johnson Space Center, Kennedy Space Center, Marshall Space Flight Center, and NASA Ames Research Center; and universities in Canada, England, and the United States). Only a few had worked together on previous projects. For some, the first expedition was also their first time working together. As such, they were at an early stage of interpersonal familiarization with one another's communication habits, speech, and gestures, which contribute to shared meaning-making. This developed relatively quickly given the intense periods of work activity (e.g., long days and co-located meals and living facilities) during expeditions.

Their analog research included simulated and nonsimulated conditions (Cockell et al. 2019; Hughes et al 2019). The latter was directed by scientists for whom data collected at the volcanic sites were of scientific interest. To have data that could lead to publication of analysis was a constant called out factor. They simulated human beings and robots on Mars. They simulated time delay (latency) between workgroups "on-Mars" and on-Earth (no need for quotes) and adjusted latency as part of their experimental conditions. Communication transmission time duration between Earth and Mars is between 5 and 20 minutes one-way. BASALT manufactured the latency between on-Earth and "on-Mars" with the communication system used between workgroups. Telecommunications between these sites were temporarily in place at the hands of two workgroups, one from Kennedy Space Center (M. Miller et al. 2019) and another from NASA Ames (Deans et al. 2017).

COTM is an area in southeast Idaho, approximately 20 miles south from the town of Arco, where BASALTers had set up their infrastructure. For some of us, reaching Arco required flying to an airport in the state of Utah and driving north for 3½ hours. Arco, population 780 in 2016, has a claim to fame as the first city in the world to receive electricity by atomic power in 1951, even if it was brief and just enough to power a few light bulbs. Arco was small but sizeable enough for businesses to provide food and shelter to out-of-towners and traffic policing. One popular diner, Pickle's Place, gave some of us our first taste of fried pickles. Of the three locations housing BASALTers, two motels and a recreational vehicle (RV) park, the Lost River Motel's structural charms included amplifying the sound of rushing winds such as on the day I could see from my window a tornado forming in the distance.

Writing in 1924 in commemoration of the naming of COTM, R. W. Limpert described the region's craters, buttes, lava tubes, caves, ropy lava flow, vegetation, water crevices, and the shredded foot pads of his camp dog. His account was part of his effort to describe and name the area in such a way as to attract tourists to an area he described as only tread by animals, fur trappers, and "Indians" (who he described in the past tense as having left trails, monuments, and markers, although there is reason to question that grammar). His interests, captured in the following line (1924, 303), coincidentally provide a segue to the second of BASALT's analog sites: "Almost entirely unknown at the present, this section is destined some day [*sic*] to attract tourists from all America, for its lava flows are as interesting as Vesuvius, Mauna Loa, or Kilauea."

A fuller view of an analog work environment includes the on-Earth infrastructure that supports the temporary apparatuses carried in by analog researchers. In Idaho, this included COTM, the National Park Service, the town of Arco, and the Mountain View RV Park and Restaurant, where BASALT had parked an RV occupied as lodging and a mobile workspace (a large trailer with a main room, smaller room, and a closet). The site's restaurant served breakfast; served as the morning meeting workspace; and prepared lunch and dinner, which were brought to a tented area adjacent to the mobile workspace.

BASALT's second Mars analog work environment in the HVNP included the field sites of Mauna Ulu, an eastern rift zone of the Kīlauea volcano (an active volcano), and the Kīlauea Military Camp (KMC), which accommodated (and fed) the BASALT team and on-Earth and "on-Mars" workspaces. Built in 1916, the camp, with its troubled past (Apple 1987; Hoshida 1981), was approximately a 1-hour drive from the Hilo airport. Although I do not write here about the third expedition, it too utilized the KMC and the Mars-analog field of Kīlauea Iki and the Kīlauea Caldera Region on the Kīlauea volcano. Approximately 5 months after BASALT finished its final expedition, the Mars-analog area collapsed in an eruption that the U.S. Geological Survey (Kīlauea 2018) called the most destructive eruption in Hawai'i in 200 years.

Kīlauea 2016, Another Interplanetary Work Environment Analog

During BASALT's November 2016 expedition at Kīlauea, hawai'i, approximately 40 BASALTers assembled to conduct daily extravehicular activities (EVAs), a term used to describe the activities undertaken by an astronaut in space and on the outside of the space vehicle. Their work system for conducting scientific investigations employed a division into three workgroups that were distributed across three worksites, one on Earth and two "on-Mars." BASALT's workflow involved distributed teams cooperatively carrying out an established set of tasks along a shared timetable for each 4- to 6-hour EVA, developed and evaluated by BASALTers (Beaton et al. 2017).

To begin to see BASALT's test setup for human–robotic exploration "on-Mars" as distributed workgroup communication, picture three workgroups—two "on-Mars" and one on-Earth. (For a detailed view with graphics, see Lim et al. 2019.) The on-Earth scientists were co-located in arranged seating in a single room referred to using a number of takes on Mission Control Center (MCC) and landing on Mission Support Center (MSC) (Figure 6.1). In 2016 Hawai'i, at KMC, the MSC was very large, with a stone fireplace and furnished with large wooden tables and chairs. The two teams "on-Mars" were a field team, which physically traversed volcanic terrain carrying backpacks with telecommunication equipment, provisions, tools, and accompanied by humans standing in for robots to carry tools and samples, and a field support crew, which was stationary and sheltered at computer consoles providing support to the traversing field crew and coordinating communication with the MCC. The building housing the MSC also accommodated the field support on console "on-Mars" team. Although in the same building, persons in this small room did not have line-of-sight to one another and could not hear each other with doors closed and headsets used by the members of the field support team.

To say this in acronym-speak, BASALT's ST (science team) were seated in the MSC in KMC from which they worked remotely "on-Mars" via audio and video telecommunication connection with two workgroups conducting the EVAs "on-Mars," the IV (intra-vehicular) crew remained stationary at consoles and the EV (extra-vehicular) crew traversed outdoors in hot, bright sunlit volcanic terrain. The IV crew were responsible for managing communication between the ST and the EV and managing resources to support the EV. The instrument suite in use and on trial was carried by the EV team.

Figure 6.1 BASALT Science and Telecommunication workgroups seated in the Mission Support Center, Kīlauea, Hawai'i, November 2016.

Although the workgroup on-Earth was labeled the ST, the majority of BASALTers were scientists by profession, with varying levels of experience. The configuration of a group of physical scientists on Earth providing support for human exploration off-Earth, for NASA, hails back to the Apollo missions for which there were scientists co-located at Johnson Space Center (in that era named the Manned Spaceflight Center). Because there were some BASALTers rotating through the crews, taking turns as EV, IV, and ST, the influence of various scientific backgrounds was a subtle (informally recognized but not formally acknowledged) aspect of data collection in the field.

A fourth team was the crew that placed and maintained the telecommunication relay technologies. Their telecommunication apparatus had to withstand natural and local infrastructure conditions on-Earth and "on-Mars." Constantly enabling the core of the temporary structure, they supported fairly constant uninterrupted information relays between persons on the volcano and persons in the camp, or even between persons in adjacent rooms (M. Miller et al. 2019; Siebert 2019).

BASALT ST produced data collection decisions that the field teams needed to progress through the EVA, according to a process designed to bring a greater number of persons and expertise to shape what was happening in the field. The EVA crew was limited to two persons, but with the ST effectively communicating with them, the number of persons conferring on data collection was closer to 20.

ST sent prioritized requests for data collection to the IV and EV, with responses returned a minimum 10 minutes later and used by the ST in their deliberations on next requests. Keeping turnaround time durations tight was a goal for a number of reasons. NASA EVAs are informed by efficiency time management logic. Mars-analog field sites are extreme environments. Minimizing time spent standing in hot direct sunlight, hiking on sharp and shifting basalt rocks, is essential to the health and safety of persons in the field.

Field instruments played a significant role in this assemblage. Digital tools, a suite of field geology instruments, were subject to ongoing evaluation for use during EVAs—it cannot be assumed that instruments used by scientists are the right fit for scientists to use in an analog and/or off-Earth. BASALT included several such tools, from hammers to spectrometers. Categorically, these tools added another dimension to the language of analogs in that two types of tools are often distinguished by way of analog (e.g., a hammer, a clock with hands) and digital (e.g., a spectrometer, an electronic face timepiece). And, like humans, the field instruments were not immune to the extreme environment.

In the rest of this chapter, I focus on an instrument's trial and an interruption offered through ethnographic data that was applied during BASALT's second expedition. My engagement changed from observer to participant observation, as did my accountability on contributing to the team's goal outcomes.

Instrument Time and Communication in Kīlauea 2016

Developing handheld instruments and user protocols is an ongoing goal for equipping off-Earth missions. Instruments have been in use for Mars exploration: The mineralogy and chemistry of the Martian surface are determined by either data sets collected by spacecrafts in orbit or on the ground by rovers (Carr 1996). One example aboard the European Space Agency's Mars Express is a visible and infrared mineralogical mapping spectrometer that detects carbonate-rich regions and other minor minerals on Mars at medium resolution (Poulet et al. 2005). Another example is the chemistry and camera tool ("ChemCam") on NASA's Curiosity rover; the ChemCam takes in situ measurements on mineral type and chemical composition (Lanza et al. 2010).

User protocols for instruments have been described (Luedemann 1989) as procedures for system setup, hardware functioning, software, system shutdown, and requirements for documentation; and, more briefly, as standards for ensuring integrity of collected data (Olympus 2022). From a human–technology relationship perspective, bringing into focus multiple social, material, and cultural aspects of communication interactions between people, tools, and environment, the user protocol also meant the communication and time needed to place the instrument in situ and collect data and for data to be communicated (transmission mode and time duration) and interpreted (human and machine processing). In addition, the singular "human" is to be interpreted as plural, akin to technology. Tool use may appear to be physically operated in the hands of one person, yet its use is contingent on said person's interpretation of information from other sources, humans and technology.

Asked by one BASALTer to describe what I found interesting about tools, I answered with a question on language used for remotely directing instrument usage in an extreme environment—for example, when ST sent directions to EV on what area of a rock to apply an instrument. Was this language developing as part of an analog culture, or was it specific to expedition locations and/or analog team composition? Former NASA astronaut, and BASALTer, Professor Jeffrey A. Hoffmann replied with a description of the importance of communication, perspective, and interpretation in an example from his work preparing to repair the Hubble Space Telescope in low-Earth orbit (Hoffman et al. 1993). To practice making repairs to Hubble, he was supported by a lift operated by another person. Their physical orientations were different—in his physical position, the direction of up ("move me up") was different than for the lift controller's position because he was physically perpendicular to the controller. When asked to be moved up, he was moved sideways to the left, which was up for the controller.

Had there been time for a longer conversation, I would have described to him, as I had shared with Dr. Lim, that the analog itself was a sociotechnical apparatus of fascination to me and one that I viewed through a lens-combination of work ethnography and social studies of technology. Their analog employed an interplanetary work system that was relational to that which I experienced on the MER mission (2020) and with ocean science teams distributed on shore and in the deep sea with approximately a 2-second communication transmission time delay (Mirmalek et al. 2019). Inter- and intraplanetary work is distributed work that requires workgroups to have systemic connectivity via tools and social processes that are institutionally provided (and when they are not and people have to themselves provide for this infrastructure, I refer to this as infrastructure maintenance work [Mirmalek 2020, building from Star 1999]). Distributed work conditions are always being reshaped (with vacillating periodicity) in relation to group cultural norms, values, habits, taboos, and meaning-making practices (Baba 1999; Faithorn and Blumberg 2009; Mark et al. 2003; Nardi 2005; Ruhleder and Jordan 2001; Wasson 2012). Tools used to support communication across distances are also always changing form, for reasons including economic competition, technological determinist mindsets, shifts in understandings on work support, and cultural conflict. And then I would have added a few words on the social studies of technology, especially as it informed what I noticed in my notes during the first days of the second expedition in Hawai'i.

Social studies of technology grounds the standpoint technology use (e.g., popularity, rejection, modifications, and applications), and relationships are shaped by multiple social and material forces. In contrast to viewing a technology's status as determined by a single causal relationship (Bijker and Law 1992; Bowker and Star 1999), it is a standpoint that disrupts a common assessment that a technology fails to become adopted because it is technologically unfit, and the counter notion that a popular technology is adopted because it is simply mechanically superior to all alternatives. Bruno Latour's *Aramis, or The Love of Technology* (1996) is one of my favorite examples. Written as a work of detective fiction, it is based on his actual investigation of the demise of an electric train system, one that was abandoned in favor of another train system, in late 20th-century France. People, social processes, economics, political forces, and technology were all equally weighted subjects of his inquiry, in keeping with actor-network theory.

Finding an Instrument Competing for Time

BASALT's suite of instruments, subject to trial and evaluation, consisted of a handheld visible-near infrared (vis-NIR), a handheld X-ray fluorescence (XRF) spectrometer, and a forward-looking infrared (FLIR) camera. These instruments were all originally designed for use in laboratory conditions—in controlled work environments in which, for example, lighting conditions can be changed as needed, and the temperature and atmospheric conditions (e.g., humidity) can be controlled. In other words, it is the opposite of an extreme work environment.

During all three expeditions, BASALTer Dr. Alexander Sehlke focused on the instruments used for characterizing planetary surfaces in connection with both research on instruments used on Mars and within analogs (Sehlke et al. 2017, 2019, 2022). Given the EVA time constraints, it was important for the ST to quickly identify alteration states of basaltic rocks for directing sample collection. The EVs were expected to use an instrument to collect data and confirm reception of data by IV and ST before using another instrument, capturing further sample data, or moving on (Figure 6.2).

Simulation conditions in Hawai'i were described as similar to those that were used in Idaho, and they were managed for consistency across expeditions by BASALT's Concept of Operations team (Beaton et al. 2017, 2019). There were minor differences to the team roster, but the majority of participants were the same. The physical size of the MSC increased, which gave the team leeway to rearrange their workspace as needed and which resulted in changes to increase communication (Payler et al. 2019). A change in the instrument suite was noted but not expected to require adjusting the setup.

Of the three instruments, the XRF had not been used in Idaho. The XRF provided unique geochemical data valuable to the process of gauging which

Figure 6.2 BASALT field crew conducting an EVA, November 2016.

field rocks best met sampling and research objectives. It was anticipated that it would be the most useful of the field instruments. Although most of the BASALT ST was unfamiliar with the XRF, their decision-making workflow established a few months earlier in Idaho had included adopting new and unfamiliar instruments and instrument data. As such, it was not expected that the newness of the XRF would raise concerns. However, after only 3 days of use, some people considered excluding it entirely.

The initial reception of the XRF was favorable. Each expedition began with teams setting up workspaces, a few days of running through EVA elements, followed by the official set of Mission Days (MD) counting EVAs and ending with a couple of days for review, planning, and dismantling the temporary workspaces. In the first couple of days, BASALT ST showed interest in the XRF instrument's data. By the third day, the XRF was still considered "good." The XRF's capacity to measure chemical composition of the rocks and their alteration states was well received. The notion that XRF was useful, however, shifted quickly.

On the fourth day of instrument use, some STs were saying that the XRF "takes too long." Comments were made that it was cumbersome, more of an impediment than a capability. The rejecting comments were strong enough to warrant setting the new instrument aside, an early end to its trial time. As I had been observing instrument data use, transmission, and interpretation in Idaho and Hawai'i, I had not seen the tool fail (e.g., to not power on).

If there was no operational failure, then what warranted shortening its full trial period? With ongoing use, the EV could use the XRF to produce spectrometer readings in the field that were sent to the ST to evaluate and send the EV their sample selection, which would be collected and "sent" to them post-EVA. The suite of instruments had not been designated as in competition with one another. But there was a shift in the discourse, and handheld instrument use was becoming a matter of selection. The XRF could prove to be an instrument of choice, or it could be evaluated as the least useful.

Offering my analysis on these conditions was in keeping with my ongoing work practice of sharing analysis with the community of study, without expectation that it would be taken up, appreciated, agreed with. This was, and is, a process more nuanced than a document dump; it includes keeping time in mind and in action for discussing observations with participants, copious note-taking and late-night memoing, deidentifying (e.g., some individuals via workgroups generalizations), and possibly producing audience-specific oral or text representations.

I shared my identification on the ST's curious drift to rejecting the XRF with instrument suite lead Dr. Sehlke (Alex) and Dr. Lim. They shared in the curiosity, and we discussed what was making the XFR appear more difficult than another instrument. With Dr. Lim's approval on adding to the work scope under Alex's charge, he and I proceeded to work together.

Going forward, each day, Alex and I had at least two dedicated discussions on observations, and we occasionally gave Dr. Lim a status update. We looked for and made slight adjustments to the workflow for instrument data collection and communication in response to what appeared to be social process issues among the ST and the EV. Each day, we evaluated those adjustments and made a choice whether or not to proceed. The first daily discussion took place in the early morning, after the team meeting on the day's schedule, and the second discussion occurred in the early evening after the team's second all-hands meeting. Alex and I worked together to coordinate daily coverage of the distributed workgroups and a daily review of notes after each EVA. We coordinated so that while Alex was in the field with the "on-Mars" workgroup and I was on-Earth, we used phone text messaging and exchanged questions and directions on communication to follow. On days when Alex was scheduled to be in the field, I would observe interactions among the science team with the person assigned in his stead on-Earth and instrument data from Alex as a member of the field crew. When we were both in the field, while Alex was in the role of field crew and using instruments for data collection, I was present to observe, taking notes and pictures, of the crew's handling of instruments, interactions, and time durations of these activities. When we were both in the MSC, on-Earth, Alex's role involved receiving instrument data from the field crew, interpreting data, and contributing his analysis to the science team discussions.

In the data account that follows, events are noted using a day count based on the total number of days that instruments were used (10).

Considering Rejection

During mission operations the first day, XRF data had been obtained in the field and had been delivered to the ST. And in this process, there were a few social and technical factors that appeared to constitute the sense of a temporal duration that was "too long." Sitting inside the MCS, the STs were updated on the timing of field events such as when the EV team reached designated exploration areas inside HVNP and when they reached a site where they would use the instrument suite. Poised to start their deliberation process, the ST waited while the EV used an instrument and sent the spectroscopic readings. Receiving these data was not temporally consistent. Sometimes receiving the information would take longer than expected because the EV had difficulty reading the instrument screens due to sunlight glare and reflection. Sometimes the sound of wind made it difficult to hear the EV team reading data aloud. To counter the audio issues, a workaround was employed—taking a picture of the XRF display and sending the picture to the ST—although even this had difficulty because of low bandwidth, which meant lower quality images that were not possible to read.

Reading data on instrument screens, however, was not new or specific to the XRF. It was experienced and identified as an issue in Idaho. Bright sunlight made it difficult to see data on digital information display screens. Their inability to quickly read the information translated into additional time expended and unplanned for fatigue from physical exposure, cognitive processing, squinting, and so on. A few workarounds were attempted. A rubber brim was fashioned around the instrument screen to cast a shadow on the display, but to no avail. The EV team took to adjusting themselves to cast a shadow on the screen, for reading or taking a picture. Having observed this issue across instruments, I took to noting the length of time spent grappling with reading out the instrument data on the XRF and the vis-NIR. I noted there was no significant difference between instruments.

BASALT's extreme work environment involved both natural and human-produced conditions. Natural weather conditions were critical features shaping work conducted during EVAs. During the Idaho expedition, the climate was hot and dry. These conditions sometimes caused overheating of the scientific equipment and malfunctions. These conditions could affect the EV team; carrying the weight of the instruments was a human factor consideration (physiology was a subject of study; Hill et al. 2019). In the field in Hawai'i, the wet climate and variable air quality produced by active outgassing from fissure systems throughout the volcano were causes for concern because the instruments were not entirely waterproof and could be damaged through interaction with the acidic environment. Likewise, for the humans, handling instruments in these conditions was markedly different from the normative lab conditions these tools were typically used in. Daily and hourly weather monitoring of some of the environmental parameters, such as temperature, windspeed, and humidity, was critical.

Collecting data with instruments in the field was only part of the total temporal duration for getting information from the ground (i.e., scientific data from the volcanic terrain) to the ST for deliberation. I observed that another set of activities—receiving the instrument data in the MSC and having it translated or interpreted into the ST deliberation—had to be considered. After using the XRF to collect rock data at a potential sample site, these data were communicated to the ST. This information was immediately recorded by an intelligent transportation system (ITS) into the shared data management system. Whereas a knowledge of chemistry was sufficient for understanding XRF information, understanding changes in chemistry from one potential sample location to another, and determining which of these locations fulfills the research/sample goals of the particular mission day, required a level of expert knowledge based in experience with using the particular instrument.

The ST members were aware of their interpretive limits with XRF data, in varying degrees. Some paused their own reviewing and looked to Alex or another ST member for another level of analysis. Some ST members continued without pause. Some found the XRF data insufficient for supporting choices in sample collection. This workflow difference among the ST members shaped their evaluation on how well the XRF was working.

On the fifth day, during the evening review meeting, the ST requested two interpretation sheets, "cheat sheets"—one for the XRF and one for the vis-NIR. Two scientists, including Alex, were asked to provide these sheets for the next day's operations. At the next morning's breakfast team meeting, Alex described the cheat sheets that he had produced over night and emailed to the BASALTers. These sheets contained information on how spectral and compositional data correlated to the types of basalt alteration and how they connected. A few scientists asked for further identifiers: "What flags light our eyes up?" "What is the optimal spot size for the vis-NIR?" They were, in effect, asking for instrument interpretations translated into visual representations. This workaround, however, fell short. Indeed, a cheat for expert knowledge was at odds with communicating the complexity of spectroscopic data interpretation that cannot be abbreviated into visual symbols of dichotomous shortcuts.

The cheat sheet request was itself an indication that there was something different about the ST deliberations in Hawai'i. The cheat discussion provided further indication of communication issues in the human–technology relationship between some of the ST members and their instruments. For feedback on how the cheat sheets shaped the ST's relationship to the instruments, Alex added two open-ended questions to the prompts used for the ST's day-end review meetings. The prompts asked for "vis-NIR likes and dislikes" and "XRF likes and dislikes." Later, in the daily end-of-day mission review meeting, which followed the ST day-end review meeting, ST members described their likes and dislikes for each instrument. Among the responses, the XRF and the vis-NIR were given the same favorable comment that they each worked "faster today than yesterday." Alex and I reviewed this comment in relation to other operational factors: The XRF was "faster" when the scientist operating it knew to activate the 5-minute warm-up before the instrument was cycled in for use. Both instruments received favorable comments on specific scientific observations the ST members were able to make with the instruments' spectral data.

On day 6, while reviewing ethnographic data on Hawai'i and Idaho ST instrument relationships and the cheat sheet, a theme emerged regarding the absence of instrument representation. Specifically, the instrument with no human representation was treated differently than the other instruments. XRF was new to the ST, with the exception of one scientist, but even for him, using it for ST deliberation was a new experience. In Idaho, the instrument scientist had been a constant presence in the MSC and kept pace and tone with all of the other strong voices in ST deliberations. She provided consistent expertise on instrument data interpretation as appropriate and necessary. This aspect of her participation had not been recognized such that it was not accommodated for when she was unable to participate in the Hawai'i expedition.

Discussing this with Alex, he asked for another instrument scientist to be scheduled to work in the MSC while Alex was in the field, and he instructed them to provide unsolicited instrument interpretations during or after he recorded data into BASALTers data management software (xGDS; Deans et al. 2017). This was different from the previous practice of waiting to be called upon. He was directed to participate out of sync with the ST's normative communication. For the rest of the expedition, the instrument scientists made this a regular part of the work practice during ST deliberations.

On day 7, the instrument scientist in the MSC read the XRF data aloud as he recorded it in xGDS followed by an unsolicited vocal interpretation of data in relation to the ST science interests at the field site. The ST noticed. In their write-up for the day-end meeting, they stated, "Actively communicated XRF results to team, and XRF data were compared to vis-NIR results, which helped to come to conclusions as to whether basalt was altered or not." Their list of likes for the XRF included "quantitative, trustworthy, includes uncertainty, facilitates/confirms decisions." The dislikes included "requires specialist use, very small spot size can create unrepresentative rock, doesn't auto-transmit data, not idiot-proof." The XRF, unlike the other instruments, required certified training. In the field each day, there was only one member of the EV team certified to wield the instrument, which meant that the other EV could not assist in holding or using it.

ST comments about the XRF screenshots as unreadable continued: "It's too dark to see the numbers." Although this was an accurate description of the screenshots, to focus on screenshots as the source of XRF data was to ignore the fact that the instrument's data were being read aloud over telecom, recorded in the shared data management software, and could be repeated by the ITS, whose trained ear could catch and record XRF data faster than a screenshot could be captured in the field, sent, and received by the ST in the MSC. Still, one ST described that they would spend some time that evening figuring out a better way to take pictures of the XRF screen; the next day, they reported that they had no success.

On day 8, ITS continued the practice of calling out the instrument data and orally describing XRF data interpretations. Alex asked the other instrument scientist to continue this work practice, without explaining the particular reason why. Later that evening, an instrument scientist, for whom this was their first day participating in this new work practice, described with excitement a greater degree of interaction with the ST deliberation process. Yet, during ST deliberations that day, there were complaints from ST about using the XRF. I took a look back at the day's events, during which some ST members called out the XRF as "taking too long," to shed some light on these different perspectives.

Prior to an ST member declaring to the team that the XRF "takes too long," the noise level of the room was high. There was enough cacophony of multiple conversations in the ST workspace that the XRF scan data coming over audio telecom was quite difficult to hear. Some were not aware it was coming at all. In addition, there were interactions between the EV and IV prior to the ST receiving XRF data that contributed to delayed reception of data. The instrument scientist in position to record these data into xGDS asked another person (whose role was to input deliberation data into xGDS) for confirmation of what he had heard and for what he could not hear. Compared with the day before (e.g., "thanks" and "can you repeat that"), when he did read aloud XRF data, there was little or no direct acknowledgment. Undeterred, the instrument scientist maintained their work practice and did, at times, receive questions to repeat statements.

On day 9, some ST comments about the XRF taking too long had gained traction; these comments were not refuted by anyone aloud. To qualify or counter this discourse later at the review meeting, Alex and I began synchronously recording time durations for instrument use in the field and for reception of data in the MSC, coordinating our efforts via text messages. Our notes showed that the "takes too long" evaluation was not supported by data. The use of the XRF that day was not measurably different from any other day, during which its data had been successfully transmitted and used in the ST deliberations. The XRF discussion that night resulted in the development of a new feature to support ST reception of XRF data from the field (using xGDS), rather than reducing the instrument's role.

On day 10, the XRF was in full use. Although it did have some use issues, it remained consistently in use for obtaining scientific data in the field for the remotely located ST. The transmission of data directly from the field to the xGDS was operating, but it was still necessary for the instrument scientist to confirm instrument data and to provide further interpretation. In the day-end review meeting, ST's instrument evaluations for the XRF confirmed the importance of particular data details it produced and the new capability of viewing data on xGDS. Alex added a prompt to the review meeting agenda asking ST to "rank vis-NIR, FLIR, XRF by most to least useful in the decision-making process for today's sampling objectives." ST responded with ranking the XRF for two objectives: "For 'syn-emplacement' (alteration of basalt during eruption and while it was emplaced): vis-NIR, XRF, FLIR; For 'Unaltered' (basalt in its chemical and mineralogical composition unchanged since its eruption and emplacement): XRF, vis-NIR, FLIR." Across review meetings, the XRF ranked either first or second.

Conclusion

All of the instruments had some issues. All of the instrument screens were difficult to read in the bright sunlight. And the clarity of all of the verbal readouts of instrument data was subject to the natural (e.g., wind) and technical (e.g., telecommunication) work environment. BASALT had not set out to compare the two instruments to one another. Yet, through the actions taken to request which instrument to used first, for example, it was visible that there was a favoring of one instrument over another. Less than halfway through the Hawai'i expedition timeline, ethnographic data showed that the science team on-Earth was disengaging from the instrument, not because its accuracy was in doubt but, rather, because their relationship with it was too taxing. It was implicitly juxtaposed with another instrument, the vis-NIR, with which the team had a positive relationship during the earlier expedition in Idaho.

Coming to Hawai'i, the XRF had no relationship yet with the ST and their interplanetary work system. Adding it to the instrument suite was not just a matter of one more device in the rotation. I question how the relationship might have grown if the XRF had been formally introduced with features of time and communication, of which there are more than a few but I offer three: (1) How long it takes each instance of using the instrument to collect data in the field given field conditions; (2) how long it takes for said data to be sent, digitally or orally, from one site to another given the personnel and communication conditions at each site; and (3) how long the interpretation process takes for the expert and for the team to whom the analysis must be communicated for the process to be complete (in one direction).

Going to Hawai'i, I had no expectation that I would find a case of instrument rejection playing out in a manner that called to mind the ways in which representation mattered even for technology that produced its own speech. It took until the second expedition for me to feel the space afforded by an analog that stood in stark contrast with how it felt to be in a mission space that required following mission-critical-only communication.

Acknowledgments

I express my appreciation for Dr. Darlene Lim and Dr. Alexander Sehlke demonstrating an openness to views outside of those with which they were most familiar and possessing the curiosity to examine and re-examine the culture of knowledge production on- and off-Earth. I acknowledge luck and other spatial factors for seating me next to Dr. Jordan Hill both in Idaho and Hawai'i; my thanks to her for taking the time to talk through some experiences. Following the BASALT program, Dr. Lim and I worked together on an analog program for which the use of robots in the deep sea would serve as analog and experimental space for developing human and robot space exploration (Lim et al. 2021). More recently, in an unexpected return to outer space, I joined Dr. Lim to work on NASA's Volatiles Investigating Exploration Rover mission (Colaprete et al. 2023; Lim et al. 2024; Mirmalek et al. 2021, 2024).

7
The Restaurant at the End of the World

Aaron Parkhurst and David Jeevendrampillai

> "Here we are," continued Zaphod doggedly, "standing dead in this desolate . . ." "Five star . . ." said Trillian.
> "Restaurant," concluded Zaphod.
> "Odd isn't it?" said Ford. "Er, yeah." "Nice chandeliers though," said Trillian. They looked about themselves in bemusement.
> "It's not so much an afterlife," said Arthur, "more a sort of apres vie."
>
> —Douglas Adams (1983, 96)

Introduction

Since the first seeds of wheat, pea, maize, and onion were taken into space aboard Sputnik 4 in 1960, there have been numerous experiments on plants in space to explore photosynthesis and plant growth under different conditions. More recently, there are many life sciences and botanical research projects aboard the International Space Station (ISS) in which astronauts cultivate plants for consumption, asking questions on the feasibility and sustainability of "gardening" off-world. The gardening systems on the ISS, such as the Vegetable Production System (VEGGIE) and the Advanced Plant Habitat, have been successful in cultivating a range of edible plants, including forms of lettuce, radishes, and mustard greens. The narratives that accompany these experiments, from astronaut researchers on the ISS, space agency media feeds, and the life science researchers and laboratories on Earth that partner with the ISS, are marked by optimism for the future of humanity. The attention given to plants in space is marked by hope and hype of expanding human civilization off-world; the positive effects of greenery and growing plants for psychological well-being; and, increasingly, the immediate concerns of sustainable sources of nutrients for the Gateway and Artemis missions to the moon and beyond to Mars.

Aaron Parkhurst and David Jeevendrampillai, *The Restaurant at the End of the World*. In: *Otherwhere Ethnography*. Edited by: Istvan Praet and Perig Pitrou, Oxford University Press. © Oxford University Press (2025).
DOI: 10.1093/9780197790885.003.0008

This chapter presents ethnography with life science researchers who have been cultivating plants since 2011 aboard the ISS. It explores narratives regarding the role of plants in space that emerge from these scientists, a wider public engagement with plants on the ISS (via social media), and traces how such narratives are present in the future imaginaries of humans in space, particularly in the form of space habitat designs. We consider how the meanings that people derive from these plants represent the vitality of nature and plants as a form of life that can offer a type of surprise—something beyond the rational and quantifiable—and enable humans to reach a post-Earth utopia. In doing so, this chapter explores a more subtle, yet equally powerful sentiment of eschatology. We argue that plants and gardening on the ISS, as mediators of future-making, are objects and practices marked by irony and contradiction, characterized by contesting claims of utopia and apocalypse. We argue that this dichotomy is predicated upon dual figures of modernity, which advocate a bifurcation of nature and culture and the associated rise of planetary system thinking whereby the Earth's ecology is seen as something to be discovered, managed, and controlled. Such an understanding, we argue, enables a vision and rationale for a post-Earth future where life can be exported from the planet and yet remain vital and adaptive through the surprising potential of plants. The presence of plants in space, then, has multiple roles. They are studied as part of an integral life support system for sustaining off-world living, but they simultaneously assert a form of more-than-human life through the promise of vitality for a form of humanity that can thrive at the end of the world. Borrowing from anthropological and material culture analytics on utopia, science, and the Anthropocene, the chapter extends beyond the context of outer space to ask: As the Earth heats and ecological catastrophe abounds, what are the emergent forms of plant–human relations at the end of times, and what is being consumed at the restaurant at the end of the world?

Space Gardens

By 1975, the USSR had sent the first man and woman into Earth orbit, and the U.S. Apollo missions had demonstrated the ability of humans to land on the moon. Alongside achievements in space exploration, people began thinking though how society might thrive off-world, creating visions of a permanent intergenerational space-based humanity. In 1974 and 1975, there were a number of conferences at Stanford University that teased out ideas and visions for future permanent human colonies in space. One of the key organizers was Princeton professor of physics, Gerard K. O'Neil. In 1976, O'Neil published a now-famous text *The High Frontier: Human Colonies in Space*. This book set out not only reasons why humans should establish permanent colonies in space but also how this might be achieved at a technical and infrastructural level. It opens with a lengthy exegesis of a problem facing humanity—that of limited resources for an increasing planetary population. O'Neil cited multiple United Nations reports on *The Limits to Growth* (Meadows et al. 1972) that outlined the challenges of increasing human populations, energy demand, and resource use. For O'Neil, the issue of how to maintain growth can be solved through technological innovation and by moving large-scale human activities off the Earth. The book goes on to outline the detail and vision of permanent human settlements in space. These are described as cylindrical settlements that rotate to produce artificial gravity, use mirrors to channel the sun, and draw on the resources of asteroids and other planetary bodies for natural resources. O'Neil colonies, as they came to be known, have been excellently illustrated in the work of artists Rick Guidice and Don Davies. As we have outlined elsewhere (Jeevendrampillai and Parkhurst 2021), the "Stanford torus," as it is popularly known, has become hugely influential in the popular imaginary of what future space living could be like, from its use in science fiction films to its influence in Silicon Valley architecture. Common in all of these visions is the plush and plentiful amount of green space. O'Neil describes gardens and the practice of gardening throughout his text, both in technical detail and via the imagined letters of torus inhabitants that he uses as a literary device to provide the reader with a sense of quotidian habitat living. O'Neil describes how gardens are important for food, recycling of oxygen, and psychological well-being.

The ideal of the original torus has been carried forward by new space imaginaries. Blue Origin owner Jeff Bezos, for example, describes his vision of a future space settlement: "These are really pleasant places to live" (Bezos, 2019). Blue Origins' renderings feature elks on cliff tops overlooking a plush valley of a river running through a green landscape. Here, the images have a certain quality of wilderness about them, similar to those depicted by the Hudson River school of art that depicted the early U.S. western frontier as a wilderness to be conquered by European colonists. However, architectural theorist and space scholar Fred Scharmen (2019a), speaking to *Bloomberg CityLab* magazine, had some choice words for these designs:

> In Bezos's imagination, older, more future-forward buildings become parodied, privatized, and zombified. The spaces are drenched in thick rays of light, as if to preserve all of this architecture in a giant jar of treacly honey. It's not just the imagery that's stale. The framing and assumptions behind the whole enterprise are outdated, too. From the viewpoint of 2019, the simple optimism of 1975 seems quaint. .

Scharmen's critique begins a dialogue on the latent "staleness" of these verdant ideals, showing contrasting sentiments of rot embedded in utopian visions. We argue such visions are always based in a dialectical bind with notions of an imminent apocalypse.

Indeed, artist and scholar Joseph Popper takes these critiques further. He outlines how these visions are critiqued by those who see such infrastructural imaginaries as a continuation of a social order of control, where the use of resources maintains ideologies of resource extraction and capitalist logics via technological innovation and frontierism (see Popper and Berry 2021, 110). Popper cites the work of Wendell Berry, who states the project is "an ideal solution to the moral dilemma of all those in this society who cannot face the necessities of meaningful change." These people are also "the chief beneficiaries of the forces that have produced our crisis" (Berry 1977, 36). Popper situates the Blue Origin images with the dialectic of utopian space living and the impending specter of the apocalyptic forces of climate change. The paradox of utopian space visions is that they do little to critique the underlying logic of extractivism, and rather move such logics off the Earth to space-based frontiers.

Although these colonies may have had a significant influence on the popular imaginary of space station living, the designs also find purchase in the design proposals for future space bases that are currently in the advanced stages of planning. Blue Origin features designs of a future planned commercial space station named Orbital Reef with bountiful plants growing from the walls of the station (Figure 7.1).

Architects who envisage space stations on Mars or the moon frequently project plants and greenhouses as critical elements to their design renderings. In this way, space plants and gardens have long been associated with utopian ideals of future space living. Plants in this setting offer a synecdoche

Figure 7.1 Microgravity greenhouse as part of the promotional images for Blue Origin's Orbital Reef, 2021.
Source: Blue Origin.

for thriving off-world. This transition from "surviving" to "thriving" is one that the architects have emphasized to us as a key motivation in habitat design (Figure 7.2).

However, just as Popper and Scharmen have outlined the tension between utopian space living and capitalist ethos, Natasha Myers has demonstrated the role that gardening takes in orientating relations that people have between themselves and plants, where plants stand as a figure of nature. In her analysis of large-scale urban garden infrastructure, Myers (2019) asks,

> What might we learn about plants and their people by examining the aesthetics and politics of garden infrastructures? And how, in a time of massive

Figure 7.2 Mars habitat living quarters submitted as part of a NASA centennial competition by Hassell Studio. Hassell's images of Martian bases are replete with plentiful decorative green plants and views into greenhouses.
Source: Hassell Studios.

> ecological destruction, are people renegotiating their relationships with plants? How are people designing gardens to stage plant–people relations otherwise? (116)

With these questions in mind, we can turn to the cultivation of plants aboard the ISS, and public response to such plants in space, in order to analyze the ways in which plants illuminate the emergent configurations of the future.

Capsicum and Mustard Greens

There are mustard greens, radish plants, chili peppers, and romaine lettuce in lower Earth orbit. They are grown in chambers scattered across different international modules of the ISS—Columbus, Kibo, and Destiny. The plant chambers, with their fluorescent lights and green leaves, are aesthetically striking against the white wire-laden technological chaos of the ISS's interior. They are fitted with blue and red LED lights that utilize the right energy and wavelengths to maximize growth efficiency. The high energy of the blue lights encourages leaves to grow, and when combined with red lights, the LEDs encourage flowering. The result is often a vivid fuchsia, radiating a sultry glow and creating shadows against the suspended equipment of the research modules. As a result, they cannot be missed when they are open and active. The neon light is not entirely out of place in the Hephaestian backdrop of the research vessel. There is irony, perhaps, in the need for LED lights in lower earth orbit to provide the energy plants need to grow. In principle, the ISS has access to unfiltered solar light to which nowhere on Earth has access.[1] It could be argued that the LED lights provide "unnatural" energy for an "unnatural" environment, although anthropologically speaking, what constitutes as "natural" is a red herring in understanding life off-world. The idea of naturality is riddled in the crisis of relativism and crouched in a terrestrial knowledge system incommensurable and inadequate (and perhaps irrelevant) to living off-Earth (Olson 2018). To begin to use a word such as "natural" to describe the cultivation of life on the ISS, or Artemis/Gateway, the moon, Mars, and wherever the cosmic asymptote leads, requires first a critique and grounding of provinciality and commonality to build a language the discipline has yet to theorize and is perhaps still unwilling to prepare to (see Chapter 4, this volume). But, ethnographically speaking, it is not "naturality" that the astronaut gardeners and the Earthbound botanists who govern them are necessarily after—its efficiency and eventually sustainability. Research on plants aboard the ISS affords a new type of scientific control beyond the scientific botanical conventions of light, water, and nutrients; it offers a botanical science of gravity as a vector that alters plant biology, and when partnered with light and other environmental factors, it can manipulate the yields of plants. Recent space flight experiments have shown that there is no radical alterity to the morphology of higher plants grown on the ISS in either the long or short term, when the environment is otherwise controlled. Plants still grow toward the direction of light, and in line with water gradients, as they generally would on Earth. However, vegetable cultivation is still an emerging practice in microgravity environments, and there is much still to learn on how the challenges of off-Earth farming influence plant growth on a cellular level (Wolff et al. 2014). For our interlocutors, the contribution to scientific knowledge was clear. However, as science and technology studies (STS) scholar Paula Castaño has shown, the ways in which research platforms in space, plants included, are valued and evaluated by the international space industry, market economies, and governments are complex and indeterminate, and they warrant critical and ethical scrutiny (Castaño 2021). Yet still, plants offer things to the cosmic gardeners beyond the immediate contributions to biological sciences. The plants are memento vitae of Earth when Earth's absence is otherwise felt. The effect on astronauts extends beyond sentimentality—the plants' materiality is visceral. The leaves are a texture of life that can be felt and tasted, and they can bring joy.

As we wrote this chapter, European Space Agency (ESA) astronaut Thomas Pesquet and Roscosmos Cosmonaut Pyotr Dubrov had just tasted the first harvest of chili peppers grown in space. The media images show it as a dark green, perfectly manicured capsicum. The pepper is unique for current space plants because of the fastidiousness it requires, its relative delicacy, and its resource intensity. After consuming it, they stated,

> The very tip did not contain any bitterness at all, but the closer to the petiole, the hotter, and the seeds were just fiery! Maybe it's the peculiarity of this sort, or maybe it's the peculiarity of growing in zero gravity. (Dubrov 2021)
>
> I didn't anticipate how good it was to see and smell, and feel plants up here, something alive in the middle of all this cold technology. I miss our Earth, and all the wonderful food it has to offer! (ESA 2021)

Pesquet notes that the plants offer something more than a science experiment. There is a sense of extraness from the plant life, as he sets them against "the cold technology" and notes how the Earth has these "wonderful" offerings. Here, implicit is the potentiality of plants to offer something wonderful, perhaps surprising. Such sentiments are echoed in the words of other astronauts regarding their interactions with plants in space.

Plants on the ISS exist in multiple social registers. They are a pride of scientific enterprise, a constant reminder of human ecology and what one needs in order to live, a source of nutrients, and perhaps even a source of psychological well-being. They are synonymous with the planet, a way to gaze back at Earth, and, for some, indexical of ethical commitments between individuals, and between one and the planet. Indeed, many astronauts have gazed back at the Earth's physical landscapes and its capacity to host and propagate life as part of their dedicated missions to better humanity. British astronaut Piers Sellers by education was an ecologist and meteorologist; K. Megan McArthur, onboard the ISS as we wrote this chapter, is an oceanographer. New National Aeronautics and Space Administration (NASA) recruits include Zena Cardman, a geobiologist. They advocate for climate awareness, and they also grow mustard greens in space. The plants also unexpectedly offer some respite from other duties. As life sciences researchers and astronauts regularly tell us, the crew often becomes enamored with the plants. The astronaut scientists spend significant research time working on their own bodies, for exercise and experimentation, and they spend significant time in maintenance and mission work, so working with plants becomes, for some, enjoyable rather than scheduled labor.

VEGGIE launched from Kennedy Space Center onboard a SpaceX commercial resupply mission in April 2014. It was installed in ESA's Columbus module 2 weeks later by astronauts Rick Mastracchio and Steve Swanson. Following installation, researchers began to notice that astronauts often worked with VEGGIE in their own time. For the astronauts, the fruits of this labor bring joy. As Jessica Mier (2019) writes,

> Did you know we're growing Mizuna lettuce on [the ISS]? We'll need to be more sustainable and grow our own food for future, longer-duration missions to destinations like Artemis and Mars. Plants also have a positive effect on our psychological well-being (growing green things makes me happy!) So excited for the harvest in the weeks to come . . . I eat salad every day on Earth, but it has been 29 days now without one!

This joy is extended to space-plant researchers on Earth, those scientists in charge of monitoring and micromanaging the vital aspects of plant growth. Our interlocutors spoke often of their love of science, and excitement for the potential developments in their field that the ISS offered. We conducted a series of interviews with an interdisciplinary team of plant scientists working on experiments involving plant seed germination aboard the ISS with whom Jeevendrampillai had been working for a number of years. Most were honest that although they have been working in space science for more than a decade, they are not really space enthusiasts (and could take it or leave it). Their passion lies with botany, and space, they say, simply offers an environment to experiment differently and, to a degree, away from the constraints of having to justify research in terms of commercial applications. Space offers an environment in which the wondrous potential of plants can be explored as "science for science's sake." The different environmental conditions of being able to fully control light and explore the gravitational aspects of plant growth were one unique aspect of the space environment, but so were the geopolitical aspects of space as a place. Our interlocutors explained the ironic conditions of how funding for research that pushes the boundaries of botanical scientific knowledge is, to a degree, more attainable for space-based projects than for similar research on Earth. One scientist suggested that much of their research could, perhaps, be conducted on Earth, but it was less likely to be funded. The international cooperation and relative geopolitical neutrality found in space meant that "science for science's" sake could receive successful funding. Another scientist, Sarah, an engineer who designs growth chambers, outlines "on the moon, energy is crazy expensive, so you need to optimize the tech, this pushes you . . . you don't think about the return of investment in space, you make it happen." Space was a place where the plant "came first, not the human" because outer space has the potential to open up the surprise and wonder that are imminent within plants. In this regard, the researchers admitted that they were not particularly space enthusiasts. And yet, the image of their plants growing in microgravity did offer moments of existential wonder that surprised them. As Karina notes,

> I must be careful not to give life to other people's feelings but mine. But we saw our work in the launch—saw it being installed in the ISS—we had quad screens set up, with six video feeds on one screen. Two of them could see the blue earth sliding by behind the plants. In the end, that plant didn't grow actually, but you could see your stuff actually in space, the plants above the planet. And it gave me goosebumps.

These researchers have been sending experiments to the ISS every year for more than a decade. Their interests in plants are diverse, leading to developments in the science and application of control rooms, the relationships between gravity and light on plant yields, and to engineering of closed-loop biological systems. They emphasized that much of what they have been able to accomplish was not really about human space futures but, rather, about the Earth. Many of the closed-loop systems the scientists were developing demanded extreme use of resources and minimal energy input. Their science opened up new forms of technology to advance agricultural production in places where it is otherwise difficult to grow plants. As one of the biologists said of space, "If you can grow plants here you can grow them anywhere." However, this form of belief in the technological ability to maintain and expand the limits to human resource use is also marked by pessimism for political change. As one interlocutor noted, "Developing a greenhouse on the moon is much easier than changing people here."

For many scientists, their calm and stoic demeanors are often betrayed when they speak about the idiosyncrasies of their work. Changes in plant yields that occur with a slight change in centrifugal stimulation or LED shift thrill the Earth-bound botanists. However, the fact that the plant grows in space at all is met with ambivalence. The director of the research center is often surprised by the attention that their work receives: "It is odd, for some I think it's about imagining otherwise. This research starts opening new perspectives of what is possible. Some people can relate better to plants in space than they can to humans." This ability to relate to plants in space, and their ability to hold a capacity for excitement for future visions of human–plant relations, also maps out to the social media sphere. Fresh food aboard the ISS is extravagant. In terms of nutritional value, leaves and vegetables are, at the moment, little more than garnishes for prepackaged, travel-ready meals. Yet Instagram posts of leaves and plant life aboard the ISS have a huge reach. Usually, interest in space activity is limited to major events, such as a launch, a landing, or a docking. Plant research is not so much an event but a process. However, the eating of a plant allows a specific moment of public attention.

With regard to outer space activity, related media is usually rife with the languages of the extreme, the amazing feats of engineering, and the wonder of life that can allow plants to grow in space. However, is the fact that plants grow on the ISS all that remarkable? The fact that plants grow *at all* is incredible. Plants operate metabolic processes by turning carbon dioxide into sugars, fueled almost entirely by sunlight. Water and nutrients help, of course, but plants produce the most complex forms from the most basic things. They are nothing short of extraordinary. So, when plants are grown on the ISS, they too are extraordinary, but not more so than if they are grown on Earth. Indeed, they grow better in the conditions they are given on the ISS than they would in the average British home, and the astronaut researchers who take care of them are probably far more adept at keeping them alive than one might be with their houseplants. Photosynthesis works, and mustard greens grow, or it fails, and the plant dies (Figure 7.3).

When media, space agencies, astronaut Instagram pages, and popular science magazines become excited about space, we suggest they are less excited about the plant that grows, but of the range of futurisms that it implies. There is first a reflection on the mundane, that the seemingly quotidian—lettuce growing—is made profoundly extreme. But there is also the promise of what the plant can give to move humans through the cosmos. As Jessica Mier (2019) states above, and as is often stated in discussions on plants in

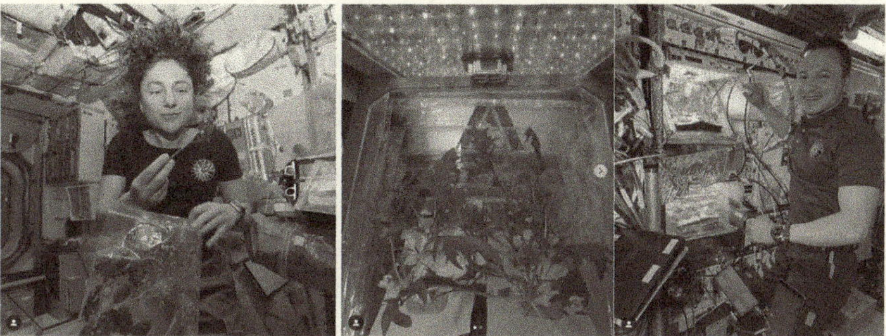

Figure 7.3 (Left) NASA astronaut Jessica Meir dines on fresh Mizuna mustard greens she harvested aboard the ISS. (Center) VEGGIE plant system in the Columbus Module. (Right) Andrew Morgan gardening on the ISS, working with The Veg-04B experiment.
Source: NASA.

space, the plants first move us to the moon and Mars. From there, human expansion into the solar system and beyond becomes tangible. It is also, as so many anthropologists, STS scholars, and other social scientists are quick, and we suggest correct, to criticize, a propagation of the old but so very hardy binarism of nature versus culture. The sentiment is that it is an extreme example of the human's mastery of nature. The plants seem to shout, then, "How ultimately modern is the ISS?"

Plants as Systems/Systems as Exportable Edens

To understand the role of plants in the visions of a space-based humanity, one needs to think through plants in two related ways. First, one needs to critically think about the ways in which plants are understood through the logic of what Valerie Olson (2018) has outlined as systems thinking. That is, they are both understood as a biological and ecological system to be examined, understood, altered, and developed. Furthermore, plants are seen as part of a wider socio/bio/technological milieu of a totally enclosed inhabitable environment in which they are involved in a relational dynamic with all other forms of life in that habitat. However, we also argue that further to understanding the position of plants, as part of a logic of emancipatory systems that free human settlements from the Earth, plants offer something more. Plants offer something beyond that which can be revealed through an increased understanding of their role in a system. It is their very ability to reveal *new* aspects of a system, *new* affordances and possibilities that gives them an imminent potential for future configurations of life. These affordances enable the expansion of humanity into space to be understood as an endeavor that will see the continuation of discovery and the expansion of the frontier. However, as Myers (2019) has shown in her analysis of urban green spaces in Singapore and elsewhere, highly technological solutions provide bountiful and lush green spaces that are seen as vital in the face of ecological collapse. Thus, future frontiers are pivoted by a sentiment of ruin, where the "Edenic Apocalypse is the embodiment of 'proper' relations between humans and the natural world, relations explicitly conditioned by capitalism and colonialism" (143). Borrowing Myer's language, we argue that plants in space and space gardens provide their own "Edenic Apocalypse," presenting human and planetary utopias in the age of anthropocenic collapse and climate crisis.

Space Plants, Capitalism, and Ruin

We adapt the title of this chapter from *The Restaurant at the End of the Universe*, conceived by British author and satirist, Douglas Adams (1983). Adams' restaurant, Milliways, is built atop the ruins of a dying world and used as a fine dining establishment to witness the end of creation (before desert is served). It is, ultimately, an ironic satire on death, as Adams writes, "Life is wasted on the living" (35). There is other social commentary in Adams' text—one can meet their dinner before it is killed and cooked (and have meaningful conversations with it first)—and it is worth noting that the restaurant is described as having extraordinary cocktails. Striking is the image of wealthy diners being so very excited and enthusiastic over the end of creation, watching the universe collapse onto itself in a kaleidoscope of colors while they sip expensive drinks. The ways in which the restaurant operates break most systems of reason, so Adams had written Milliways a slogan: "If you've done six impossible things this morning, why not round it off with breakfast at Milliways, the Restaurant at the End of the Universe." Fresh food aboard the ISS is not nearly as extravagant. Leaves and vegetables are, at the moment, hardly more than garnishes for prepackaged, travel-ready meals, although they are marked with a similar sense of wonder and impossibility. A cynical observer may claim that the veggies might better be understood as social media fodder for the resurgence of space enthusiasm. Their potency lies less in the present and more in potentiality and future-making.

Within the space industry, Olson (2018) notes how her interlocutors at NASA describe how developing a systems-based approach enabled them to shift from one problem to another. Charles, a civil servant who does thermal systems work for the ISS, tells Olson, "As a systems engineer I like knowing I can apply all my tools to anything, even human beings" (23). Here, systems thinking is a foundational epistemological approach that enables all things to be seen with a wider context of relationality with their environment. This epistemological approach to environments can be scaled from a focus on microecologies or to relations at the scale of planets and beyond. The "system" as a mobilizing concept of thought and knowledge has been fundamental in the understanding of the planet in the age of human-influenced climate change. The changes in climatic conditions of the Earth, namely its radical heating, are understood within a system in which humans, specifically those who are entangled in the systems of economics and social orders that result in the burning of carbon on a huge scale, can affect the planet to such an extent that they will make a profound mark on its geology. Dipesh Chakrabarty proposes in *The Climate of History in a Planetary Age* (2021) that "anthropogenic explanations of climate change spell the collapse of the age-old humanist distinction—prevalent in the seventeenth century but dominant really in the nineteenth—between natural history and human history" (26), namely that of nature and culture. Chakrabarty builds on Bruno Latour's (2012) thesis that human society "has never been modern," whereby Latour argues, in his anthropology of science, that much of modernity is a matter of faith. Latour argues that distinctions between nature and society, between nonhuman and human, that arose within the modern, post-Enlightenment, scientific method have masked the degree to which notions of culture and nature are particular and contingent to the ways in which knowledge is constructed through epistemic practice. As Jane Bennet notes in her book *Vibrant Matter* (2010), the nature–culture distinction provides only a thin description of the world. At a time when interactions and relations between the human and nonhuman are becoming more intense and vital, it is of little surprise that there is a proliferation of work on the more-than-human and the notion of nature–human distinctions (see Braidotti 2006; Haraway 2013).

To demonstrate systems thinking, we take the case of spirulina. Spirulina is a biomass of cyanobacteria, concentrated blue-green algae that is high in protein and many nutrients, does not require external food sources, is high in antioxidant activity, and combined with the right diets may have a range of health benefits beyond nutritional value for astronauts. It is, in principle, extremely easy to grow and can produce high yields. The potential physiological and economic benefits of spirulina are currently being researched for space exploration as well as applications on Earth (Fais et al. 2022). Figure 7.4 shows models of computer-controlled photobioreactors that demonstrate how spirulina can be grown in extreme environments, such as Mars or desert environments on Earth. The reactor provides the right amount of light, the correct mix of nutrients, and a supply of air bubbling through it. The algae is filtered out easily for harvesting. Spirulina as an example of a future technology is a microcosm of a utopian material ecology: Perfect sunlight, perfect air, perfect nutrients. Its utility on Mars is fairly obvious, but a malevolent observer will note that its need on Earth is a condemnation.

Other plant research for space futures has involved corporate purchases, private investments, and industry hype. A new product called Airocide, licensed and branded by KES Science & Technology, in partnership with Akida holdings (both American companies), is an air purifier that destroys airborne bacteria, mold, fungi, mycotoxins, viruses, volatile organic compounds such as ethylene, and odors. The device has no filters that need

Figure 7.4 Spirulina tanks and photobioreactors on display at the London Design Museum.

changing and produces no harmful byproducts, such as the ozone created by some filtration systems. Food preservation customers include supermarkets, produce distribution facilities, food processing plants, wineries, distilleries, restaurants, and large floral shops. The implications of Airocide are complex. It has important uses, perhaps in hospitals and clinics because it is highly effective in preventing the spread of bacteria and germs. It has immediate use, then, as Earth-bound medical technology, derived directly from the novel air scrubber on the ISS that controlled the environment for ADVASC mustard greens and soybean plants. Its primary use, however, is in the preservation of perishable goods in the context of overpopulation, desertification, unequal wealth distribution, and resource scarcity. It is no surprise that some of its first commercial buyers were the oil-rich nations of the Arabian Peninsula for use in refrigerated trucks and grocery stores.

The positioning of plant science in space, then, can be understood within the logics that the Earth has an increasingly fragile and vulnerable environment that is unable to sustain the human need for resources. This is reflected in the approach of those who have thought about human settlements off Earth from Gerard K. O'Neil in the 1970s to Elon Musk today. Musk, who owns rocket company SpaceX, has stated that his motivation for envisioning humans settling on Mars is that

> there is "some probability" that there will be another Dark Ages, particularly if there is a third world war. . . . We want to make sure that there's enough of a seed of human civilization somewhere else to bring civilization back, and perhaps shorten the length of the Dark Ages. Solon 2018

British astrophysicist Stephen Hawking echoed this sentiment and often referred to human endeavor in space as "life insurance" (Shiga 2008).

Examples of this approach exist on Earth in such places as the Svalbard Global Seed Vault. This can be understood as simultaneously a celebration and cathedral to plants and life and a monument to the impending threat to the ecology of the planet's global ecological collapse. Its very purpose is to save and preserve the diversity of nature in the inevitable and already ongoing processes of mass extinction and desertification. It is ironic, then, that this archive is increasingly under threat from global warming as increasing global temperatures melt the permafrost each summer, flooding the entrance and threatening the seeds in its vaults. As we have argued, the idea of having a seed archive or growing plants in space offers more than a relationship with plants based on utility. Rather, plants offer something more. Given Myers' (2019) sentiment of an "Edenic Apocalypse," we suggest that the ISS also performs an analogy for a Noahide Arc. That is, we suggest the ISS and other space environments are places where the promise of plants can be realized in the face of ecological collapse on Earth. The pragmatics of science and cultivations for these space plants for the future of humanity is already a dangerous contingency. Astronaut Don Pettit (2012) gave voice to his space plants—zucchini seedlings in his case—when he was onboard the ISS, a process that anthropologist Debborah Battaglia (2012) refers to as a form of ventriloquism, highlighting the fragility of plants and, by proxy, people in the face of public narratives of triumph and dominance. Although it hasn't the heart to tell its astronaut caretakers, Pettit's zucchini ultimately confesses to its audience that it will never bear fruit, an allegory hinting toward deep-rooted structural imbalances within global human futures—in this case, less metaphorically, gender inequality and underrepresentation. The zucchini explains its logic: "I make two kinds of flowers; male flowers with only stamens and female flowers that produce zucchinis. Being part of this all-man crew, it was fitting for me to make only male flowers" (Pettit 2012). What good is Noah's Arc if all the flowers are male?

Plants as More than a System

Plants or rather mushrooms (which are their own genus) were the central figure in Anna Tsing's (2015) engaging work, *The Mushroom at the End of the World.* Tsing begins her work with a type of apocalyptic question: "What do you do when your world starts to fall apart?" (1). Tsing begins to answer this question by going for a walk in the forest and examining mushrooms as she ponders the terrors of the world. She focuses her attention on the ways in which matsutake mushrooms emerge as a source of value in forests marked and affected by the human activities of industry, nuclear war, and deforestation. For Tsing, the matsutake is positioned as a figure of surprise and hope in a climate of apocalyptic ecological breakdown. She states,

> Terrors, of course, there are, and not just for me. The world's climate is going haywire, and industrial progress has proved much more deadly to life on earth than anyone imagined a century ago. And it's not just that I

> might fear a spurt of new disasters: I find myself without the handrails of stories that tell where everyone is going and, also, why. (1–2)

Tsing (2015) provides both a methodology and an ethnographic object to think on what thrives in the wake of human destruction. She shows what a rare mushroom can teach us about sustaining life on a fragile plane. Her methodology follows the mushroom as an object that offers surprising and complex chains of global human enterprise, industry, and relations. The figure of the mushroom situates nonhuman life as a source of optimism in the age of the Anthropocene. As she writes, "When Hiroshima was destroyed by an atomic bomb in 1945, it is said, the first living thing to emerge from the blasted landscape was a matsutake mushroom" (3). We argue that dining on mustard greens on the ISS, or the idea of plants on Mars or the moon, offers a similar optimism. This optimism is not so much that humans can understand and control ecological systems but, rather, is based on the fact that plants may offer new sources of hope despite humanity's effect on ecological systems. That is, in space, we need to take Eden, nature, and the unquantifiable element of plant life with us in order to maintain the promise of a vibrant future.

In this regard, plants offer more than research, they offer more than well-being for astronauts, and they offer more than what can be found through a system: They provide an Eden-like promise. Within the Judeo-Christian tradition, salvation and eschatology are in a binding relation. This relationship is well-illustrated here. As we have written elsewhere, some of the most celebrated thinkers in space science and engineering publicly share this view. Scholar and philosopher Emmanual Levinas, in his poetic essay "Heidegger, Gagarin and Us" (1990), echoes the arguments of Bruno Latour. Levinas writes,

> "Everything that, for centuries, seemed to us to be added to nature by man, was already shining forth in the splendours of the world. Man must be able to listen and hear and reply. . . . A little humanity distances us from nature, a great deal of humanity brings us back. (233)

As we have argued in this chapter, plants offer people narratives of utopia and salvation. What is being consumed at the end of the world is a promise of vitality and an extraction of the dominance of nature. What Levinas asks, however, is that the plant, as a proxy for nature, not offer a technology of salvation, because technology as salvation is a dangerous tautology. In Romantic fashion, what the plant offers for Levinas is both a caution and code of ethics. He finds some poetry in the tamarisk, a plant that has long thrived on salted earth, used by civilizations current and long dead to keep the desert at bay (now used in large-scale anti-desertification projects in China), and said to be planted by Abraham after he dug his first well in the Negev. He argues that embedded in the plant is humanity's obligations to itself. He tells his readers to celebrate the tamarisk for what is does, for all its affectual resonances, but to keep within the mind a warning:

> Oh! tamarisk planted by Abraham at Beer-sheba! One of the rare "individual" trees planted in the Bible, which appears in all its freshness and colour to charm the imagination in the midst of so much peregrination, across so much desert. But take care! Tamarisk is an acronym; the three letters needed to write the word in Hebrew are the initials used for Food, Drink and Shelter, three things necessary to man which man offers to man. The Earth is for that. Let us remain masters of the mystery that the earth breathes. (233)

8

"Welcome to Planet Mars"

Analog-Making in Astrobiology and Planetary Sciences

Valentina Marcheselli

Introduction

"Welcome to Planet Mars" read one of the panels at the Myvatn information center in Iceland. The small green building rose from a service area surrounded by a wide car park full of buses and off-road vehicles; a few well-equipped hikers were making a quick visit to the mini-market next door to buy provisions of water and instant food before returning to their excursions. The small information center was no more than a cube clad in corrugated metal siding, filled with a huge collection of leaflets and posters displaying Icelandic wonders and advertising tours. The astonishing natural landscapes sharply contrasted with the functional and minimalist Icelandic architecture. Next to the "Welcome to Planet Mars" board, other panels showed watercolor representations of the variety of birds and flowers populating the nearby wetlands and the rich diversity of lava concretions that can be found in the Myvatn area. Every year, thousands of tourists are attracted to this place because of its geological liveliness, producing a broad range of rare and stunning phenomena. On the opposite wall, books were displayed on a long shelf; one of them was titled *The Living Earth*. The Earth, in Iceland, seems to be no less alive than plants and flowers and humans: Less than 20 million years ago, a blink of an eye on a geological scale, Iceland was not a land at all. When the Mid-Atlantic Ridge, the fissure created by the drift of two large continental plates, met a hotspot moving east, a lively bubbling and bustling of the melted magma below the thin crust allowed it to rise up to the surface, creating the ground where we were standing.

Once back on the bus, we were told that—according to traditional beliefs—elves dwell in rocks, and thus people are strongly discouraged

Valentina Marcheselli, *"Welcome to Planet Mars"*. In: *Otherwhere Ethnography*. Edited by: Istvan Praet and Perig Pitrou, Oxford University Press. © Oxford University Press (2025). DOI: 10.1093/9780197790885.003.0009 from removing or damaging the boulders and pebbles as this would upset the householders. Other traditional mythologies claim that trolls turn into stones when hit by the sunlight. The same advice holds: Do not move rocks if you do not want to upset the trolls, who are neither the smartest nor the kindest creatures on the island. In turn, for scientists, rocks are small pieces of that world that can be moved around, magnified, deconstructed into smaller components, and tested. The group of people of which I was part surely enraged the elves and trolls: For the following 3 days, we collected samples for an astrobiology exercise;[1] as the information center panel suggested, we were using Iceland as a Martian analog.

Astrobiologists and planetary scientists aiming to characterize the surface of other planets and scout for the presence of life elsewhere in the universe today make ample use of what they call *field analogs* (or analogues). Since the Apollo missions, particular sites—such as the Meteor Crater in Arizona (Messeri 2014) or the volcanically active region of central Askja in Iceland—were used as analogs for training purposes. Astronauts heading to the Moon were taught how to recognize rocks, collect samples, and efficiently use their cumbersome gear. With the end of the Apollo program, nevertheless, this kind of training lost its purpose. It was only in the early 1990s that analog fieldwork came to the fore again. The discovery of microorganisms in places previously considered inhospitable for life (the so-called *extreme environments*) raised awareness on the resilience and adaptability of microbes, and it spread a new wave of optimism about the possibility that microbial forms of life exist on other planets. But because many of the microorganisms living in extreme environments cannot be cultured in the laboratory under "standard conditions," the study of extremophiles requires that astrobiologists engage in field campaigns to collect samples and study the complex environmental conditions of their chosen field sites. Although astrobiologists freely admit that no present terrestrial analog "mimics all properties [of a specific solar system location], they play a crucial role in preparing for different aspects of future space missions" (Martins et al. 2017, 49). The activities undertaken in analog environments inform the assessment of a planet's potential for habitability, the choice of the most appropriate life-detection techniques, the discernment of what would count as evidence for life (so-called biosignatures; see Cavalazzi and Westall 2019), the mission payload testing and validation, the landing site selection process, and so on. Astrobiologists' and planetary scientists' research questions are, obviously, space-bound, but the use of Earth as a "tool" (Martins et al. 2017) makes their research more empirical (Dick and Strick 2005): Through the exploration of analog field sites, that which was once considered the speculative study of life *elsewhere* in the universe (namely *exo*-biology) can be reframed into the study of life *in* the universe (see Kaufman 2022), starting from the "one data point of life that we know: life on Earth" (Cockell 2015, 1). Astrobiologists and planetary scientists, therefore, define analog sites as much as they are defined by their use.

Scientists today can rely on a number of analog field sites made and maintained over time, from the hyperacidic lakes in Yellowstone National Park in the United States (McKay et al. 2004; Wettergreen et al. 2005) to the Atacama Desert in Chile (McKay et al. 2004; Wettergreen et al. 2005), the hydrothermal vents in the deep ocean floor (Barge and White 2017; Russell, Murray, and Hand 2017), and the Dallol Vulcan in Ethiopia (Cavalazzi et al. 2019); the list of sites and motives for their study is in continuous development (e.g., see Preston et al. 2023). Some of these sites are considered more flexible, some uniquely positioned for a certain purpose, and others simply easier to reach and explore.[2] Despite the inherent diversity of these field sites (Marlow, Martins, and Sephton 2008; Martins et al. 2017; Fairén et al. 2010), "historically," the GSA compendium claims, "there has been an *instinctive* nature [*sic*] to use aspects of Earth to explain observations of other planets and moons" (Garry and Bleacher 2011, xi).

The aim of this chapter is to look into the apparent naturalness with which analog sites are rendered. In fact, the apparently simple claim that a certain place is Mars-like or Moon-like[3] is neither a given nor devoid of consequences. The similarity relations thus built become constitutive of the ways these environments are perceived. In fact, this chapter does not take the claim that something is Moon-like or Mars-like, or like any other celestial body, at face value. Astrobiologists and planetary scientists know all too well that the analogs they use are guided by a purpose and can only be considered valid under certain conditions. Scientists are not naive. To strengthen analogs' legitimacy as heuristics, nevertheless, the alleged "instinctive" nature of the similarity is often emphasized, and ultimately, the social process and the historical circumstances behind their assessments and choices are effaced. What this chapter resists, therefore, is the idea that these kinds of claims are only descriptive and based on features that are out there in nature. My aim, nevertheless, is not to untangle completely the connections that tie Iceland and Mars together, nor to delegitimize Iceland as a geological analog site. On the contrary, in taking the relationship between

Earth and other planetary surfaces seriously, I hope to show that analogs are the outcome of an open-ended and underdetermined social process. This is of paramount importance because claims such as that which opened this chapter do not simply describe a certain site; they put it in place, and eventually shape the present and the future of space exploration.

Analogs: From Theory to Experience

Whether in remote places or in more easily accessible locations, field analog activities are foundational to the way astrobiologists and planetary scientists come to understand their objects of interest. Analogs shape the design of space missions and the way the data collected during those missions are interpreted (Vertesi 2015). In fact, not every astrobiologist or planetary scientist engages in long and adventurous field trips: Some of them focus on computer models and simulations, and others focus on doing experiments in the laboratory with samples that other scientists have collected in the field. Nevertheless, a growing portion of those who would call themselves astrobiologists and planetary scientists have started engaging in fieldwork activities, and the resulting knowledge has been used to confirm the validity and legitimacy of what is done in other experimental spaces.[4] According to Lisa Messeri (2016), the local character of fieldwork and the cosmic framework of space science are not poles apart. On the contrary, she claims,

> a place-based orientation, rather than passively gazing at the globe from the outside, allows for an imagination of being on/within/alongside, of experiencing, the planet. This is an active relationship between subject and planet, which for planetary scientists becomes foundational to how they come to know their planet of study. (12)

Messeri destabilizes definitions of place and space (the latter is often defined as universal, abstract, and blank, whereas the former is defined as experienced and thus filled with meaning [Tuan 1977]) by emphasizing that in planetary science, place is not simply subjective or individual but, instead, an "epistemological heuristic . . . that allows for a meaningful mode of interacting with objects that physically lie outside human experience" (Messeri 2016, 15–16). What Messeri calls *planetary imagination* requires that scientists situate themselves (even if only temporarily) in a certain environment and learn how to engage with it. In doing so, the Heideggerian *being-in-the-world* and what they come to imagine as *being-on-other-worlds* mutually inform each other.

In studying the perception of environment among the Sami communities in Finnish Lapland, anthropologists Tim Ingold and Terhi Kurttila (2000) claimed that traditional knowledge is not an abstract set of precepts and beliefs that can be handed on from one generation to the next but, instead, a way of being sensitive and making sense of one's surroundings. Knowledge, therefore, is not *local* in the sense that it is proper to a community dwelling in a specific region but, rather, in the sense that relates to living, experiencing, engaging with—and therefore making—a certain *place*. This knowledge[5] cannot be separated from its context of origin because it is not knowledge *of* something, reducible to the always identical repetition of a given script, but knowledge of how to attune oneself to one's surroundings. Disentangling traditional knowledge from its context would solidify it in a set of abstract, immutable, fixed—and therefore dead—principles. According to Ingold and Kurttila, on the contrary, "tradition can be continuous without taking any fixed form; . . . there is no opposition . . . between continuity and change. Change is simply what we observe if we sample a continuous process at a number of fixed points, separated in time" (192). The kind of knowledge that makes place (the same kind of knowledge that Messeri attributes to astrobiologists and planetary scientists[6]) is thus an ongoing process. Planetary analogs, therefore, are not given once and for all. Instead of considering an analog as a static entity, we aim to understand it as the continuous *analog-making process* with which astrobiologists and planetary scientists engage.

Analogies are pervasive in scientific practice (Hofstadter and Sander 2013; Lakoff and Johnsen 2013; Holyoak and Thagard 1995); they are often considered important to the representation of scientific concepts and the communication between experts and the lay public, but this is only a small part of what analogies do in scientific labor. Analogies in fact constitute the very fabric of theory-making: According to the philosopher of science Mary Hesse, the analogies between a model (in this particular case, Iceland and its environs) and the system under study (in this chapter, Mars) provide the only effective way to search and test for new hypotheses and expand the explanatory power of a theory. Similarities and differences—or what Jones (2018) would call *dissimilarities*—between the two terms of an analogy are not fixed but, rather, objects of testing and debate; in this very process lies the predictive power of analogical reasoning (Hesse 1966). Establishing an analogy requires a continuous process of similarity judgment and negotiation.

In literature, simile, metaphor, and analogy are all figures of speech in which two elements (the system under investigation, which is referred to below as "a," and the model "b") are juxtaposed and similarities are drawn between the two. The comparison, nevertheless, is made explicit to different degrees: A simile takes the form of "a is like b"; a metaphor takes the form of "a is b" ("like" is omitted); and in the analogy, the first term is completely hidden, meaning that talking about b induces the reader to refer to a. This is precisely the power of analogies: The more they become entrenched in the way an object of analysis is understood, the more powerful and stable they become, up to the point that their intentional, creative, and productive character might become hidden from view (Stepan 1986).

In their sociological analysis of scientific knowledge, Barnes et al. (1996) emphasize that despite seeming obvious, the identification of analogies in science as a contingent action is crucial. "When it is overlooked," they claim, "the result is typically a purely formal account of modelling, which fails to grasp its purposive and goal-oriented character, and hence how it comes to be recognized as successful or unsuccessful" (108–109). No model is ever perfect, they admit, and this is what makes the agreement on what constitutes a *good enough* analogy interesting to the sociologist. "A successful model," the authors suggest, "is a pragmatic accomplishment, something which those who evaluate it take to serve their purposes" (109).

To recover the social mechanisms of the analog-making process, I propose to look at the summer school analog-making exercise in Iceland through the lens of *finitism*, a constructivist and communitarian theory of meaning (Kusch 2002). The finitist account of word–world relationship is rooted in two principles. First, the most basic way of learning what a word means is by *ostension*: The repeated pointing to instances of something, such as the category of "duck," makes the learner competent in recognizing new instances of ducks. Because each duck is different, the relationship among all the examples of duck is not one of identity but of *resemblance*. This has two consequences: (a) Each competent individual might judge the next instance differently and, therefore, there is no precise boundary between one instance being judged to be too different to a certain category or perhaps similar enough (underdetermination); and (2) each new instance of a certain category affects the way the category itself is defined (open-endedness). The second principle is that no individual by themself can ascertain whether their judgment has been made competently; it is only in encounters with other members of the community that the judgment will be identified as competent or not.

But if finitism is a theory of meaning, how can it be extended to analog-making? In fact, analog-making and concept application are deeply entrenched processes. On the one hand, the etymology of concepts can always be traced back to the association to other known phenomena that somehow present similar characteristics and are therefore considered analogous. The word cell, for example, comes from the association that Robert Hooke drew between the biological structures he first observed with the microscope and the tiny single-rooms dwelling for monks and nuns. Sometimes, Jones (2018) reminds us, "even as we move along an analogical ladder, [a concept] continues to bear traces of its early ancestry, its prefigurations" (131); analogies remain embedded in concepts and thus coevolve. Furthermore, both analogies and categories are based on a similarity judgment—that is, a process of negotiation around what counts as "similar enough," which puts into play identities, values, beliefs (including different kinds of knowledge), and cognitive authority. That which is similar enough (either the next duck's similarity to all the previous ducks we have encountered or an Icelandic environment's similarity to the Martian surface) is learned through socialization and only in a second moment becomes "instinctive." The community, then, not only plays a normative role in the single similarity judgment but also provides the very basis for this judgment through training and socialization. Continuity and change, then, are part of the very same process.

But let's insist, for a moment, on the parallel between concept application and analogy-making. In a chapter titled "Stories Against Classification," Tim Ingold (2011) criticizes the very idea of classificatory knowledge as a kind of knowledge in which "every element is slotted into place on the basis of intrinsic characteristics that are given quite independently of the context" (160). Ingold shares Bloor and colleagues' aim to emphasize the open-endedness, creativity, and contingency of any knowledgeable doing—and yet, they seem to take opposite directions. Ingold suggests that knowledge should be understood as "integrated not by fitting isolated particulars encountered here and there into a categorical framework . . . but by going around in an environment" (160). In other words, knowledge, for Ingold, is not classificatory, but *storied*, and cannot be "transmitted" from generation to generation; on the contrary, people "grow into it by following trails through a meshwork" (143), or what he calls *wayfaring*. Finitism, on the other hand, is a theory of meaning deployed in the mid-1980s to provide a foundation to a new kind of sociological approach to knowledge and knowledge-making. The so-called Edinburgh school addressed science as a social, human, and communitarian enterprise, thus shaking the idea of scientific knowledge as *universal* and *objective*, and of science as an asocial, ahistorical institution, detached from the cultural context within which it emerges and thrives. Through the principles of meaning finitism, Bloor and colleagues (1996, 46) aimed to show "the role of experience and tradition in the act of classification."

My aim is to weave a new thread in the tapestry and, together with Ingold and Kuttila, emphasize that that tradition shall be understood as a way of living, experiencing, and engaging with a certain *place*. In moving the attention from analogies as stable entities to the continuous work of analog-making, I welcome Ingold's suggestion to treat the kind of knowledge produced and reproduced in scientific practice as *wayfaring*. But instead of rejecting the idea of classification—the assessment of what is similar *enough*—or dismissing scientific knowledge-making,[7] I look into its details by engaging with analogy-making: If, on the one side, analogies reproduce the process of accounting for what counts as similar *enough*, on the other side they are rich in storied relationships. The relationality that is disclosed in the analog-making process has to be understood rather literally, not as the coupling between predefined and fixed entities but, rather, as the "retracing of a path through the terrain of lived experience" (Ingold 2011, 161). It is precisely in the lived experience that the Edinburgh school's approach to knowledge-making and the Heideggerian's attentiveness to place-making meet and are revealed as complementary frameworks to make sense of analog-making as practice. The principles of finitism are thus enriched with an embodied aspect that has too often been overlooked.[8] In this chapter, finitism becomes a thinking tool to reframe analog-making practices as communitarian, goal-oriented, under-determined, and open-ended. In turn, the telling of stories from the field helps reveal the myriad relations that are spun together in scientific practice.

(Un)Making Analogs

In this section, I look into some of the similarities and dissimilarities between (specific areas in) Iceland and (regions and epochs of) the Martian surface and subsurface through the lens of finitism. Each of the five principles of finitism is followed by a vignette from my ethnographic observations, interviews, and document analysis of Iceland as a Martian analog.

Principle 1: Future Applications of a Term Are Open-Ended → Future Analog-Making Is Open-Ended

According to this first principle, a word does not correspond to a close domain of applications but, instead, to a series of exemplars to which the word has been previously applied and which, therefore, guide how it will be applied in the next case. The word "rose," for example, is applied to a large set of flowers; when we encounter a new flower, all the previous examples of roses we have seen will guide the assessment of whether this new example can be considered a rose or not. We need to make a decision about which characteristics define a flower as a rose. The outcome of this judgment, nevertheless, cannot be anticipated with certainty: Roses can be different in size, color, number of petals, quantity of thorns on the steam, and so on. In nature, no two things are identical, and thus being under the same category means being similar *enough*. But similar with respect to which characteristics? How much is *enough*? If this decision can be considered trivial when we see a rose in our garden, it is much more troublesome when a botanist names a flower obtained by the hybridization of two different species: Would the new flower count as the same rose (perhaps with small variants) or as a new kind of rose altogether? These are issues—either the simple cases or the more complex ones—that need to be considered every time a term is applied to the next instance.

The same can be said for space analogs: There is no definite class of environments or environmental features that can be considered Mars-like (or like any other celestial body), and no closed domain in analog-making. Treating an environment as an analog field site emphasizes certain characteristics as salient while, at the same time, demanding that other features are left in the background—overlooked or simply considered insignificant (even if, perhaps, only temporarily). Listing all the differences between Iceland and Mars would surely be unnecessary: from the composition of the air we breathe to the water flowing in rivers, ejected by geysers, poured in waterfalls, and condensed in small dewdrops.[9] No place in Iceland, and in fact no place on Earth at all, has exactly the same conditions that one would find on the Red Planet: Our neighbor planet is colder and drier, with no vegetation and not enough oxygen for humans to breathe. But reaching the conclusion that Iceland cannot serve as a Martian analog misses the point of the scientific exercise.

The Martian surface is constellated by volcanic features whose alterations are interpreted as aqueous; in other words, scientists think that when Mars was still a young planet with a thicker atmosphere, it experienced intense volcanic activity, and probably layers of ice and liquid water. Because water has almost disappeared from the Martian surface in the present, the interpretation of these features is made possible by studying how similar morphologies and mineral compositions formed on Earth. This is what makes Iceland, with its 30 active volcanic systems in constant interaction with glaciers, waterflows, and permafrost, a "natural laboratory for analog studies." Iceland, in other words, represents a sort of "playground" (Martins et al. 2017, 73) for scientists because many different geological features can be found within a few hours' drive: basaltic lavas, glaciovolcanism and periglacial features, lava tubes, caves, ridges, cones, hydrothermal deposits, hot springs, geysers, and so on. Not only are these latter formations relevant on the geological level (to ascertain how present structures can be interpreted as the result of past processes) but also their being inhabited by a number of more or less complex microbial communities makes them important from an astrobiological perspective. Recently formed lava fields are studied as uninhabited habitats (Cockell 2014) that are progressively colonized by microbial populations; hot springs and geysers become the abode of extremophile communities. Extremophiles, literally "lovers of the extreme," are microbes that survive in conditions (temperature, pressure, salinity, acidity, level of radiation, etc.) that most humans and nonhumans would find prohibitive and forbidding. The term includes many different types of microbes: halophiles, literally "salt lovers"; acidophiles and alkaliphiles, whose optimal growth is at low or high levels of pH; thermophiles, thriving at temperatures above 80°C; and so on. The resilience of microorganisms has encouraged scientists to rethink the possibility of life on other planets: What might once have seemed too prohibitive has been reframed as "extreme." In fact, extreme is an interesting term because it both reflects a deeply anthropocentric perspective about what constitutes a suitable environment for life and, at the same time, shifts this perspective by acknowledging the potential commonalities between Earthly and other planetary environments (Helmreich 2006, 2012).

The formation of a riverbed and the presence of microbes in hot springs are different phenomena, but both inform research questions relating to different regions and different epochs of the planet. Driving for a few hours in Iceland, therefore, leads planetary scientists and astrobiologists to feel as if they are traveling through the geography and history of the Red Planet. In doing so, nevertheless, they purposefully ignore[10] all the differences that would break the analogy apart. Although it is not obvious, the analogy is easily recognized not only by the scientists who engage in fieldwork in Iceland but also by the rest of the astrobiology and planetary science community reached by the space analog narrative, who learn to see a bit of Mars in Iceland's mesmerizing beauty.

What is the sociological significance of this first principle applied to analog-making in planetary science? "In so far as classification is conventional," Barnes et al. (1996) claim, "it is our selected ways of applying our terms which will determine what form the conventions have and will come to have, not the conventions which will determine our ways of applying" (55–56). Astrobiologists and planetary scientists make continuous reference to new and better analogs; traversing Iceland (as much as traveling to any other analog field site) allows scientists to identify new analogies, and thus broadens the set of fields they have at their disposal for studying Mars by proxy. The analogies and similarity relations that can be drawn cannot be anticipated. Previous scientific knowledge and practice can only provide a set of exemplars; the way they will guide analog-making, even if not fortuitous, is an open-ended process. What relations will prevail and persist is revealed by the sociological study once a process is observed in its historical development.

Principle 2: No Act of Classification Is Ever Indefeasibly Correct → No Act of Analog-Making Is Ever Indefeasibly Correct

As discussed above, because analog-making does not proceed on the basis of *identity* but, rather, *resemblance*, there cannot be a unique correct way of extending an analogy. Each time the analog-making process is carried out, the social actor evaluates the new case and the way the analogy might apply to a new site, or could be extended within the same site. The level of competence in this process, nevertheless, cannot be decided by the individual who executes it in the first place. It is the community of reference who will embrace their choice as correct or reject it as misguided or unsound. These two reactions, in fact, are only the extremes of a continuum on which one can find different degrees of negotiations. Most of the time, the reconciliation of individuals' perceptual intuitions is accomplished implicitly and unproblematically; sometimes, nevertheless, the process requires a more explicit, laborious, and conflictual negotiation. "From the sociological perspective," Barnes et al. (1996, 56) wrote, "this highlights the role of collective judgements" and the way they intersect with individuals' discernment: The community both enables the individual's judgment by providing previous instances of the analog-making process and overrules the judgment when other individuals (who may bear more cognitive authority) consider it unsuited.

Learning how to engage with the analog-making labor that characterizes the astrobiology and planetary science community was the purpose of the summer school activities in which I was taking part. For this purpose, the summer school was organized into two stages. During the first few days, the group was given lectures on the geological history of Iceland, with its record of interaction between volcanism and glaciers; the climate, water, chemical, and physical conditions of the Martian surface and subsurface; and the methods of biosignature detection. The lessons provided a set of examples of analog settings and analog-making practices and negotiations. During the next days, we then undertook a practical exercise. We moved to the Myvatn area and grouped into four teams, each equipped with a set of tools.[11] Once there, we visited four different sites, where we were allowed to collect a maximum of 12 soil samples to be analyzed for the quantity of adenosine triphosphate (ATP), a molecule utilized by all (known) forms of life on Earth to store energy, and thus employed in this exercise as a proxy for the quantity of living microorganisms available in each sample.[12] The exercise, nevertheless, did not just involve a simple measurement of ATP: The goal we had been appointed was to formulate a good research question to be answered through the ATP assay. The activity was deeply rooted in the idea that—either for the reasons with which we had been supplied or for reasons we were allowed to negotiate from anew—Iceland could represent Mars. How? And what for? This was for us to assess.

The evening before we started visiting the field sites, we met in a small bungalow with our laptops and a copious amount of coffee in order to decide how to proceed. The first decision to make was what we could find out about

Mars from Iceland, given the resources and limitations we had been allocated. We eventually decided to collect rock samples from the Holuhraun lava field[13] in the Askja region. Because the erupting magma was rich in volatiles, the small bubbles coming out of the solidifying lava created a rough rubbly surface of broken lava blocks with very jagged sharp edges and a network of tiny porosities within the rocks. Microbes, we thought, might have progressively migrated into these tiny vesicles, and so we decided to assess how deep inside the microorganisms had penetrated since the recent eruption. The experiment was inspired by the idea (brought up during one of the lessons) that on Mars, microbial colonies might have migrated underground and within the rocks' fissures, thus being sheltered from the radiation that reaches the Martian surface and threatens any life forms. As both the National Aeronautics and Space Administration (NASA) and European Space Agency (ESA) had prepared for missions equipped with drilling tools, we considered this issue to be of considerable interest. The analogy we tried to establish was guided by a "purposeful" parallel, a commonality actively sought in light of the forthcoming experiments to be performed on the Red Planet.

Once in Holuhraun, however, we quickly realized that the procedure we had planned was simply not feasible. The rock was too hard to crush without crumbs and pieces being scattered around and tearing the plastic bags apart. Practicalities made us rethink our analog-making exercise, as the sample collection procedures were considered unsuited to the activity. Not without a bit of frustration, we rearranged our experiment for the Krafla region, a site characterized by hydrothermal systems; we reached the region the following day. Instead of probing the microbes' ability to colonize spaces progressively farther away from the surface, we shifted our topic and our analog-making effort to the survival of microbial colonies in the proximity of hot springs. Another group collected samples from the Hverfell crater, a tephra cone volcano whose caldera they sampled at different heights in order to evaluate how microorganisms are carried around by the wind and eventually settle on different parts of a slope. Where on the slope do microbes best survive? Another group focused, as we did in the first place, on the Holuhraun lava flow; however, they did not study pores. Instead, they used a drone to scout the area and simulate the coordination of remote sensing and field practice in the sample collection process. In the end, we all imagined a different condition to be tested and different scopes. Most important, we all made suggestions as to how the analogical reasoning on which the experiment was grounded could have been carried on. The first stage of negotiation was carried out within the small communities constituted by our own teams. After the exercise, we all discussed how well the analog-making fitted with the aims we pursued and the practicalities of the exercise. The correctness was not self-evident but instead based on these aims and practicalities.

The summer school simulated, on a smaller scale, the process of analog-making and the subsequent negotiations around assessing its adequacy and suitability, which are routinely undertaken by the astrobiology scientific community. "Intuitions"—which, as previously argued, are not an individual fact but, rather, the consequence of socialization into a certain community—would remain a speculation if the rest of the community did not recognize the similarities between Mars and the chosen Mars-like environment. The analog-making requires the community to recognize (and, as we shall see below, reconsider) its legitimacy, correctness, and suitability.

Principle 3: All Acts of Classification Are Revisable → All Acts of Analog-Making Are Revisable

If the next application of a term (or of an analogy) is open-ended and it is only in the exchange with other members of the community of reference that its correctness is assessed, then the indeterminate character of a term can also be extended back in time. Something previously considered an instance of a certain category might be reassessed as an instance of something else.

Iceland has not always been "Mars." Had we traveled to the same regions in the 1960s, we would have been welcomed to a Moon-like land instead. During the Apollo program, approximately 30 astronauts in total bounded across northern Iceland in order to learn how to recognize geological features and how to select and collect meaningful (according to geologists' and planetary scientists' judgment) rock samples. NASA had sent astronauts to places such as Hawaii and Meteor Crater in Arizona, but the basalt rocks and volcanic geology of Iceland's barren highlands were believed to be the most Moon-like of any training area. During their stay, they visited places such as the Askia volcano, the Hrossaborg Crater, and the Drekagil lava field, where they could find "beautiful volcanic geology with practically no vegetation cover. Features include calderas, ash cones, steaming volcanic vents, cinders, pumice, various types of lava flows. Probably the most moon-like of the field areas" (NASA, GFTs July 12–16, 1965 Iceland). "I spent around ten days exploring the volcanically active regions of Iceland, a place so stark and barren I felt as if I were already on the Moon" (Worden as cited in van Niekerk and Hoffmann 2019) remembered Al Worden, the command module pilot for the Apollo 15 mission. Astronauts and their trainers did not simply base their visits on mere aesthetic considerations; they were aware of the goal-orientedness of their field trips. "Being in the field at such analog sites was made to be both a legitimate and necessary way of knowing other cosmological surfaces," writes Lisa Messeri (2014, 205). "It is through these embodied fieldwork practices that astronauts refashioned the Earth as the Moon, and the Moon as Earth, so that they could make scientific sense of the lunar surface" (205).

The story had long been forgotten, until a young Icelander named Örlygur Hnefill Örlygsson found a newspaper from the 1960s showing a picture of the Apollo astronauts with the caption "They're coming!" Intrigued by the idea that astronauts basically trained to walk on the Moon in his backyard, Örlygsson started collecting anecdotes and memorabilia from people who still remembered the astronauts' sojourns. In 2012, he inaugurated The Exploration Museum, a small exhibition celebrating the Moon landing and the role Iceland played in the enterprise. Today, however, Iceland has stopped being an analog for the Moon and has become Mars instead. "The Apollo astronauts I have met," Örlygsson reported,

> have told me that Iceland [is] an even better place to train astronauts to visit Mars than it was for the Moon. . . . We know a lot about the geology of Mars because rovers have been studying the surface of Mars for some time. What they have discovered is that there are many more similarities between the geology here in Iceland and the geology of Mars. (cited in van Niekerk and Hoffmann 2019)

Mars and the Moon are two radically different places—but this did not prevent people from acknowledging certain similarities with the Moon in the past, and then revising their judgment and focusing on other similarities that lead it to resemble Mars instead. The Moon analogy has not been openly rejected but, rather, gradually substituted by an analogy that better fits with the present interests. This makes the process all the more interesting; the fact that both Mars and the Moon, despite their many differences, were considered analogs to Iceland, and that the shift from one to the other did not pass through a moment of contestation but followed the path of space exploration and the gradual shift of interest from the Moon to Mars, makes the contextual, social, and historical dimension on which analog-making is grounded more apparent.

Principle 4: Successive Applications of a Term Are Not Independent → Successive Applications of Analogy Are Not Independent

Unlike astrobiologists and planetary scientists, who travel to places such as Iceland and other analog field sites in order to produce new knowledge about space, Apollo astronauts traveled to Iceland in order to train for what they would do once on the Moon, and how to do it (what instruments to use, what samples to collect, how to collect them properly, etc.). This latter use of space analogs is usually labeled "simulation." Analog sites can be used either as a source of environmental information or as an arena to optimize the efficiency of an instrument that will be part of a rover payload (or improve one's skills with a certain tool, as in the case of the Apollo astronauts). Often, analog field sites are used for both. A planetary scientist and leading member of the Mars Society phrased it as follows: "In my view there are four applications, or four uses, of analogs: One is to learn, one is to test, one is to train, then the fourth one is to evaluate, or to reach out." Learning refers to the use of a field as a geological analog that can in turn be used as a proxy for another planetary environment in terms of geology and biology. "While we're learning about Devon Island," he explains, "we're learning about Mars. It's quite fundamental."[14] Testing refers to the assessment of instruments' functioning, operability, and efficiency. This is because when pieces of technology are brought to the field, many unexpected circumstances unfold: Dust might clog a gear, remote controls might work intermittently, wheels might struggle on irregular, muddy, or inclined surfaces, and so on—the list could be limitless.

Training refers to both the practices of using these instruments and working in conditions that simulate those on another planet. This was also the case with the Apollo astronauts who trained for the Moon landing in the same regions that are used today as Mars analogs. Astronauts were trained in field geology: how to recognize geological features, how to select a significant sample, and how to carry it back safely and without contamination. However, although manned space missions are not yet planned (they are a long-term goal), training remains today a central aspect. One of the trainers of the summer school put it in an apparently different way: Analogs can be about science, technology or *exploration*. In other words, analogs can be used to learn something about Mars, how to use a certain tool on Mars, or how to experience it. Exploration and explorers embody the analogy and enable it to shift from one level of experience to another. Many (perhaps most) analogs share this multilayered structure. The major space agencies periodically update their catalogs of analog environments, specifying their best uses—and most catalogs offer the entire set of possible uses. The assessment of the composition of a rock, the calibration of a rover instrument, and the sense of loneliness one might experience on a barren and remote land are, nevertheless, very different things. Through analog-making, nevertheless, they cannot but become one and reinforce each other. It does not matter that they correspond not only to different activities but also to different moments in history (the rock formed billions of years ago, the technological instrument is in the present or in the near future, and the problems connected to the psychology of living on Mars are far in the future). How is it that they can pertain (without apparent contradiction) to the same embodied activity?

Logistics often impose that the different people involved join forces; sometimes this is simply because the co-occurrence of different objectives allows funding to be more easily granted given the higher probability of achieving at least one of the declared objectives. Some people are involved in projects tackling different aspects of space missions and thus cannot help but carry with them their entire set of research interests.[15] Sometimes field expeditions are organized in regions that have often been used and are, therefore, better characterized; this means that at least part of the complex work of getting to know a field site has already been accomplished, and thus different tasks might be carried out more easily. Talking about the Atacama Desert, a planetary scientist I interviewed claimed that

> apparently... it's been visited by a fair amount of people, because it's started to get crowded. You cannot walk without seeing a rover, whether it's from NASA, in which case it would be mine, or it's a rover from ESA, or well... the whole place is populated by our instruments and robots.

She has herself used the Chilean *altiplano* to make an argument in favor of her favorite landing site on Mars, and also as a place to test a technology that would allow a mission to land on and explore Titan (one of Saturn's satellites).

Every time astrobiologists, planetary scientists, and astronauts think through an analog and start presenting their research, issuing reports, submitting projects, refining skills, and justifying experiments through that analog, the rest of the community's practice of analog-making cannot remain unaffected. This is even more crucial given the interdisciplinary and multipurpose use of analogs. In other words, when a certain place is made Mars-like, what it means to be Mars-like shifts slightly, and therefore this action is liable to condition all subsequent uses of the expression. Analog practices are shared as part of the epistemic culture of the collective so that the way one uses the expression "Mars-like" will affect what another will do with it. As Barnes et al. (1996, 58) suggest, "From a sociological perspective, the interdependence of different acts of classification is a way of describing a form of interdependence between people."

Principle 5: The Applications of Different Terms Are Not Independent of Each Other → The Applications of Different Analogies Are Not Independent of Each Other

To be fully competent in the analog-making process, an individual must be competent in the use of the whole system of classification and narratives. In the same way that successive applications of an analogy modify the extent of that analogy, they also modify its relationship with associated terms, categories, narratives, and meanings. On a sociological level, "developing interactions between the uses of terms will be identifiable empirically as developing interactions between people" (Barnes et al. 1996, 59). These interactions are not abstract but, rather, situated and entrenched in social actors' contexts; they also shape possibilities for future actions and connections.

Örlygur Hnefill Örlygsson, the young founder and director of The Exploration Museum, was born in 1982: "long after the last man walked on the moon," he emphasized. Encouraged by one of the Apollo astronauts who had returned to Iceland in the early 2000s, Örlygur decided to collect all the documents and testimonies still available—pictures, letters, postcards, personal items, and training plans—and to display them in order to celebrate the role his small country played in what he defines as the greatest achievement of humankind. The exhibition grew into what is today The Exploration Museum, a place where Vikings, polar adventurers, and astronauts are all considered to be different embodiments of the same ethos: exploration. The museum "just happened to be here, in Husavik," which was the first Viking settlement in Iceland, founded in 870 AD. "The connection was already there," says Örlygur.

These associations are not neutral; they do not leave the idea of *exploration*, through which so many of these enterprises are represented and carried out, unchanged. Strongly influenced by the NASA missions, The Exploration Museum is not immune to the references to the U.S. narrative; however, within the Icelandic framework, the events celebrated acquire new nuances. The thread that unites the three major exhibitions (the Viking sagas, the conquest of the poles, and space exploration) is an omnipresent reference to the breaking of human limits. The museum is infused with a sense of progress, ecumenism, and essentialism. The aim, claims the museum's website, is "to show how the story of exploration is our story, what makes human beings human, because exploration is not just the discovery of new lands, it is mainly the discovery of new things about ourselves, about humankind in general" (The Exploration Museum n.d.). And yet, there are also novel elements, which can complement other narratives and perhaps allow us to glimpse alternative directions for the future. First, the Icelandic space chronicles—with their roots in the local specificities of Husavic, the northern permanent settlement in Iceland, within the Arctic Circle—present the narrative from the periphery. The perspective of Husavic is too northbound to appeal to the typical Western-centric viewpoint; for instance, the museum's story does not refer to the American frontier as the archetypical movement of conquest and settlement. Second, the museum collection gives a sense of human connection through small gestures, such as the reciprocal exchange of postcards between Neil Armstrong and his Icelandic hosts, rather than mere celebration of epic enterprises. There is also something deeply humbling in the Icelandic stories, a sense of struggle and occasional defeat. Many Viking expeditions were not successful, and the North Pole explorers' biggest success was being able to return home.

These narratives, and the analogy that Örlygsson established between Armstrong and Leif Erikson (the Viking explorer who, according to the Icelandic Saga, first discovered North America), between going to the Moon and going to the South Pole, between the old space race and the contemporary missions, between the Moon and Mars, and so on, allow new threads to be weaved and do not leave the analog-making process unaffected. Taking seriously how making a place Mars-like works locally in the many locales where the analog process takes place helps contrast the universalizing narrative of space exploration and keeps the pace with space exploration's always shifting meaning.

Conclusion

Most environments can be used, in a way or another, as space analogs—mines, caves, volcanoes, deserts, the ocean floor, Antarctica, or even a department roof (Cockell, Hecht, and Landenmark 2018). All of these places can be turned into Mars-like sites, and never for an obvious, self-evident, or indefeasibly correct reason, but instead because of the way the community embraces each case as legitimate and (at least temporarily) correct. Space analogs are naturalized—that is, are made to seem natural, objective, and given—by the narratives often adopted to describe them. The "Welcome to Planet Mars" that opened this chapter, together with the many other similar expressions that can be found in astrobiological discourse, objectifies the relationship between the Earth and Mars—or, shall we say, between particular locales on Earth and on Mars. But far from being inscribed in nature, analogs are continually created and sustained through the experience of them (Marcheselli 2022). By focusing on the analog-making process, this chapter aims at relating the continuous assessment of what counts as *similar enough* to the storied world within which this activity always and inevitably takes place. The five principles of finitism that have been used as a thinking tool in this chapter to follow the many threads that are weaved in the analog-making process had been formulated—after all—as a tool to emphasize the open-endedness and creativity implied in any act of classification; by embracing Ingold's approach to knowledge as wayfaring, the chapter emphasized how exploration of analog environments continuously allows an ongoing negotiation of identities and the retelling of stories about the world we inhabit. Moving beyond the apparent opposition between a purported classificatory vocation of scientific knowledge and a more creative way of knowing, I claimed that every act of classification, of meaning attribution, of analogy-making *is* a way of telling and retelling stories—stories of the past and stories about the future.

Astrobiologists' and planetary scientists' vision of Iceland is not neutral compared to the traditional mythological framework, in which rocks are the materialization of elves and trolls. These two visions are comparable and allow for different kinds of engagement with the environment. Although in some cases setting up and exploring analog environments allow for the protection of a territory (Cockell 2007, 121; Gunnarsson and Örlygsson 2019), they also introduce new vulnerabilities and normalizes certain behaviors and actions (Mendenhall 2018; see also Chinigò 2019). This is why the continuous and detailed study of analog-making is important: It shapes not only the kind of environment we experience and engage with on Earth but also the kind of approaches and attitudes that are considered possible, desirable, or inevitable in other planetary environments.

9

Artifacts of Attunement

Victor Buchli

Introduction

The material culture of low Earth orbits is characterized by a body of material culture that although conceived and mostly built under the conditions of 1 g on Earth is used and occupied under the conditions of microgravity in low Earth orbit (LEO). Thus, this body of material culture functions mostly under alien contexts unknown on Earth yet materially and functionally is a product of both 1 g and microgravity. In these respects, the material culture of LEO accommodates two radically incompatible and exclusive domains of human existence, at once of the conventional conditions of 1 g and the extreme and alien conditions from a terrestrial perspective of microgravity. This investigation is prompted by an observation made by Alice Gorman in regard to her archeological work with Justin Walsh on the International Space Station (ISS). Gorman (2017) observed that a category of artifacts are present on the ISS that serve as "gravity surrogates." That is, such artifacts serve to mimic in some fashion key aspects of 1 g to sustain life in microgravity. These vary from Velcro to straps, plastic Ziploc bags, duct tape, exercise equipment, and the bodily orientations of the inhabitants of the ISS that attempt to mimic the conditions of 1 g in the microgravity of LEO.

It is the contention here that this emergent body of material culture, although unique in that it attempts to embody two radically incompatible realms of human existence, 1 g and microgravity, nonetheless highlights enduring controversies in the anthropological characterization of material culture that challenge how we understand material culture not only on Earth but also in contexts such as LEO and others.

That a single artifact or body of artifacts might embody two radically incompatible realms is a long-standing problematic within material culture

Victor Buchli, *Artifacts of Attunement*. In: *Otherwhere Ethnography*. Edited by: Istvan Praet and Perig Pitrou, Oxford University Press. © Oxford University Press (2025). DOI: 10.1093/9780197790885.003.0010 studies. From the multiplicity of meanings associated with material artifacts, and its radially contextual nature to "material culture" as analytical category developed within Western typologies and intellectual traditions, to the role of things and artifacts of various other contexts which are radically distinct from this master frame of analysis that characterizes material culture itself and its analysis (Strathern 1990). Whereas such analyses focus on how realms of meaning are radically incommensurate, within a realm of conflicted interests, the issue I bring to bear here is how such incommensurate realms are made commensurate not just in terms of radical excess and open-ended "bundling" (Keane 2005, 2007), which has been shown to characterize these incompatibilities radiating outwards so to speak, but also in terms of how these realms are actually made—improbably—to be commensurate and the implications of such tensions that inhere therein for our wider understanding of material culture tout court. It is the contention here that unlike approaches which highlight the excessive and unstable nature of any given artifact and their contextual realms of stabilization, the material culture that emerges in LEO extraterrestrially must attempt to make the radically and materially incommensurate realms of 1 g and 0 g in fact commensurate.

Pietz and the Fetish

Here, Pietz's conceptualization of the fetish is highly instructive. His historical account of the first encounters off the west African coast in the 16th century that produced the "fetish" is an account of how to attempt to stabilize radically incommensurate realms. This is a conceptual problem; I contend that it is not dissimilar to those encountered in LEO. I use two artifacts from Pietz's discussion—West African gold weights and Portuguese monumental Padrao—and contrast them with two examples from LEO on the ISS, namely Ziploc bags (following Gorman) and one the first three-dimensional (3D) printed tools in LEO (Made in Space 2020)—a wrench printed in 2016 on the ISS (see Figure 9.3). All four artifacts I propose share certain similarities that are useful to think through in relation to Pietz's discussion of the fetish that emerged as a result of the clash of European and African encounters off the coast of West Africa during the 16th and 17th centuries in a context of a novel and radical incommensurability that can be usefully contrasted, I contend, with the radical incommensurabilities that emerged 400 km above Earth in LEO in the first part of the 21st century. These four artifacts in turn

I position following Graham Harman's insights derived from Heidegger's lectures from the War Emergency Semester of 1919 (Heidegger 2008) in his development of an object-oriented ontology. In short, I talk about a weight, a pillar, a plastic bag, and a wrench and their radical incommensurabilities as they are territorialized in time and space and different conditions of gravity.

The West Coast of Africa

As mentioned previously, Pietz traces the emergence of the concept and the artifact category itself, the fetish, as a territorially and historically unique occurrence during the 16th and 17th centuries specifically along the Mina Coast of West Africa (Pietz 1985, 5–6), where the fetish "brought together previously heterogenous elements into a novel identity" (7).

The fetish is characterized as a "trifling" by Pietz (1985, 9) from the European perspective, a point I return to later when considering rather banal artifacts such as "clips" and "Ziploc plastic bags" in LEO. Here, seemingly banal artifacts from a European perspective achieve a novel distinction as a "fetish" within this encounter between two incommensurate realms. As a result of this encounter, as Pietz observes, "In Marxist terms, one might say the fetish is situated in the space of cultural revolution, as the place where the truth of the object as fetish is revealed" (11). I refer to this point later when discussing what is revealed by such "artifacts of attunement" in LEO, such as wrenches and plastic bags, but also how it might highlight our understandings of material culture in terrestrial contexts as well.

Furthermore, the fetish, according to Pietz (1985), is characterized following Leiris by "the sheer incommensurable togetherness of the living existence of the personal self and the living otherness of the material world" (12). In particular, Pietz observes,

> This object is "territorialized" in material space (an earthly matrix), whether in the form of a geographical locality, a marked site on the surface of the human body, or a medium of inscription or configuration defined by some portable or wearable thing. The historical object is territorialized in the form of a "reification": some thing (*meuble*) or shape whose status is that of a self-contained entity identifiable within the territory. (12)

Here in Pietz's observation, the issue of territoriality, radical singularity, and the intimate embodied nature of the fetish draws attention to the ways in which such radical incommensurabilities negotiated by such artifacts of attunement in the context of extraterrestrial environments function to radically resituate the human body.

Furthermore, Pietz (1985, 13) notes that "each fetish is a singular articulated identification (an 'Appropriation,' *Ereigenes*, in Heidegger's language) unifying events, places, things and people, and then returning them to their separate spheres (temporal occurrence, terrestrial space, social being, and personal existence)." Here, the radical singularity of the fetish also exemplifies its radical dynamism—once localized, it is dispersed again. And contra Heidegger, who in his later work presents this "gathering" in the rather static terms of the fourfold, the discussion here explores the decidedly tense and unstable terms of the fetish as Harman (2019) is very keen to point out apropos the fourfold that resonate closely with Pietz's own characterization.

But more important, in terms of the way in which our concepts of territoriality are configured, Pietz (1985, 15) observes that "the fetish is precisely *not* a material signifier referring beyond itself, but acts as a material space gathering an otherwise unconnected multiplicity into the unity of its enduring singularity—the category of 'territorialisation' was established.". These qualities I contend are particularly present when considering the artifacts of attunement I propose that emerge in the context of LEO and the novel conditions of a territoriality and its space–time that emerge therein.

Pietz's (1985) argument focuses on a particular category of artifacts, namely Akan gold weights: "The gold weights, then functioned precisely to relate incommensurable social values, those from traditional Akan culture as expressed in proverbs or traditional healing, with the newer market values introduced from outside" (16). Such gold weights regulated newly emergent relations of capitalist trade and simultaneously incorporate the intimate well-being of the human body that emerges in those encounters. But more relevant to the discussion here, Pietz unwittingly introduces the value of 1 g as key to the territorialization of the fetish, fixed as it were under the conditions of Earth gravity and thereby uniting disparate elements under the conditions of a particular universal terrestrial force of gravity that ensures a stability of weight across terrestrial space–time, and in fact establishes a unified territory of exchange according to emergent merchant capitalism and European cartography and navigation, where one unit of weight strives to be the same and interchangeable in a unifying sphere of emergent global capitalism that harnessing and universalizing the workings of 1 g secures as well as the intimacies and health of the human body in a radical and novel configuration.

Inherent to the fetish is its relation to the human body and its sustenance:

> The active reaction of the fetish object to the living body of an individual: a kind of external controlling organ directed by powers outside the affected person's will, the fetish represents a subversion of the ideal of the autonomously determined self. (Pietz 1987, 23)

Similar to the extraterrestrial contexts such as LEO, the skin-enclosed Western Cartesian self is controlled and connected through various prostheses, intrusive measurements, vast networks of control and management, and artificial atmospheres, without which it could not be "natural"—that is, viable or sustainable (see Valentine 2017b). The human body in extraterrestrial contexts is a "natural" cyborg that challenges received Western ontologies of selfhood and autonomy despite seemingly valorizing them to be only inverted under extreme conditions extraterrestrially. The extreme as discussed by Battaglia and others works to in fact assert this deeper condition that is also at the heart of many non-Euro-American ontologies (Battaglia et al. 2012).

Furthermore, echoes of a cyborg-like nature resonate in Pietz's discussion: "Thus fetishes were external objects whose religious power consisted of their status almost as personal organs affecting the health and concrete life of the individual" (Pietz 1987, 44). For the 16th- and 17th-century European, this observation confirms the "irrationality" of newly encountered peoples of the coast of West Africa, whereas in the 20th- and 21st-century context of LEO, it is most emphatically the techno-scientific "rational" condition of the human body extraterrestrially.

I consider here another fetish discussed by Pietz in addition to the Akan gold weights: the monumental posts erected by Portuguese authorities, the Portuguese Padrão. Both "fetish" artifacts create two distinctive yet related territorializations: the gold weight in terms of the interchangeability of incommensurable elements within a nascent European capitalist exchange sphere under the conditions of 1 g, and the European navigational coordinates of the cardinal coordinates of north, south, east, and west. Here, as Pietz (1985) notes with regard to the Padrão,

> The side facing north bore the arms of the royal house of Portugal; the west face situated the moment of erection in time reckoned in relation to the death of Christ; the south face situated the moment of the pillar's fixing

> in time of the reign of John II; and the east side declare the act of fixing the pillar in place to be the deed of the Portuguese noble Diogo Cão (17),

thus coordinating these disparate realms in a novel territorialization, recognized as a fetish, locally by people encountering the posts on the ground. In both instances, radical forms of territorialization are achieved attuning in what might be described as an agonistic tension between these disparate elements producing two distinctive yet related novel time spaces at the heart of early capitalist and colonial exploration and expansion.

Attunement and Tension

As opposed to recent conventional approaches within anthropological studies of material culture that destabilize, liberate, or render excessive in terms of figurative excess, or vibrant artifacts of material culture studies, artifacts of attunement seem to function differently. This seems to be a process of the "extreme" conditions of a given body of material culture, whether off the western coast of Africa or 400 km above the earth, where what might be construed as excessive realms of material experience are made commensurate in terms of a radical attunement, rather than a radical excess. One might consider this a form of "gathering" (to use Heidegger's language) in a centripetal fashion as opposed to the dynamic and open-ended dispersal outwards characteristic of recent approaches that emphasize radical indeterminacy and excess—which in turn might be considered centrifugally and as is present in some discourse—as "liberatory." The situated and singular tension inherent within the fetish as described by Pietz is highly informative here as to how this "gathering" in more static Heideggerian terms as Harman (2009) observes is fraught with lines of tension, dynamism, and barely contained instability.

Centripetal and Centrifugal Dynamics

Contrary to more conventional approaches that aim to liberate, radiate, or destabilize the artifact of material culture per se, here an opposite dynamic is proposed that characterizes the peculiar status of such objects reminiscent of Pietz's fetish, that reconcile 1 g with 0 g/0.166 g/0.376 g be it in LEO, the Moon, or Mars, respectively. I suggest that an internal and radically localized and singular tension inheres between a centrifugal and centripetal dynamic, that each in its way consolidates meaning and materiality along different axes of centrifugal or centripetal motion, where both movements equally attempt to singularize a decidedly unstable dynamic whether centrifugally or centripetally as a consolidation in relation to one pole or another as Harman (2009) suggests apropos Heidegger's four points of the fourfold. This can be characterized as the expansive consolidation toward an imperial center—the classical object of material culture par excellence in its traditional museological nexus as exemplified in the Pitt-Rivers famous chart of 1868 (Figure 9.1). Or this might be seen in an opposing fashion in terms of the consolidation in the opposite direction toward a non-Western, local, and recontextualized center. Here, Taussig's (1993) discussion of Darwin's observation of such first encounters and clashes between the inhabitants of Tierra del Fuego and first European contact is instructive. Darwin observed how the resolutely egalitarian Cuna would tear up a European man's shirt into individual pieces to be distributed outwardly and equally among them

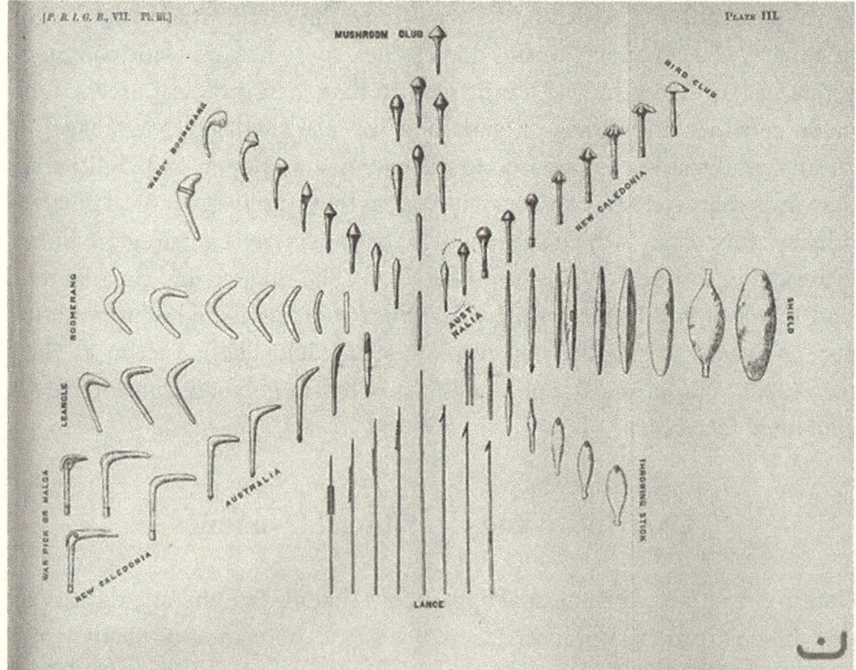

Figure 9.1 Pitt-Rivers famous 1868 plate, illustrating the evolution of culture.
Source: Pitt Rivers Museum, University of Oxford.

rather than to be worn as befitting a singular and individuated European person (137–139). Here, an egalitarian centrifugal dynamic disarticulating and distributing outwardly is counterpoised to the hierarchical individuated and centripetal dynamic which consolidates inwardly to focus on the production of the Euro-American individuated self—that something like "one's own shirt" affords. Here, something avowedly and "imperially" universal as Euro-American individuated selfhood is rendered parochial and local within an asymmetrical although radically distinct sphere, where nonetheless both contexts are distinctive and mutually exclusive dynamically—the difference between a singular shirt (centripetal) and the violently multiple "trifling" shreds of cloth (at least from the perspective of a perturbed European) distributed outwardly (centrifugally). The result of which is that the "shirt" itself cannot be held together, cannot "gather" to use Heidegger's language, and literally breaks up into numerous shreds of cloth. The work here is in the productive restructuring of this dynamic which produces localized knowledge with its practices and the work—if you will—of appropriation on either side of asymmetrical structures of power that produce contingent forms of being.

It is worthwhile to reconsider Pitt-Rivers' famous chart of the artifacts of "primitive warfare" (see Figure 9.1), which reverses the centrifugal multiplicity of encountered forms of weaponry across the world in space and time in relation to the centripetal forces that converge on the "*ur*" weapon at center that embodies the "virtual" truth of an artifacts essence (following Willerslev 2011). This chart (or cosmogram as Tresch [2007)] would have it) can be opposed to a similar "cosmogram"—the map produced by Made In Space of one its 3D prints in LEO (Figure 9.2)—and the orbital paths of a given print run across the planet which affirm the locality of the artifact as being in fact "made in space" while its manufacture traverses multiple time zones, nation-states, and borders—never made in the United States, China, or Russia, nor Britain, despite the co-temporal timing of its manufacture in Greenwich Mean Time. Both cosmograms as befit the term (see Tresch 2007, 2012) are totalizing schema of territoriality erasing long-established localities toward the constitution of a transcendent center (both virtual and physical) and with this an evolutionary and neo-evolutionary teleology: in the case of Pitt-Rivers, the 19th-century scheme of unilineal social evolution, and Made In Space's own claim to create the conditions of a new multiplanetary civilization in the not too distant future.

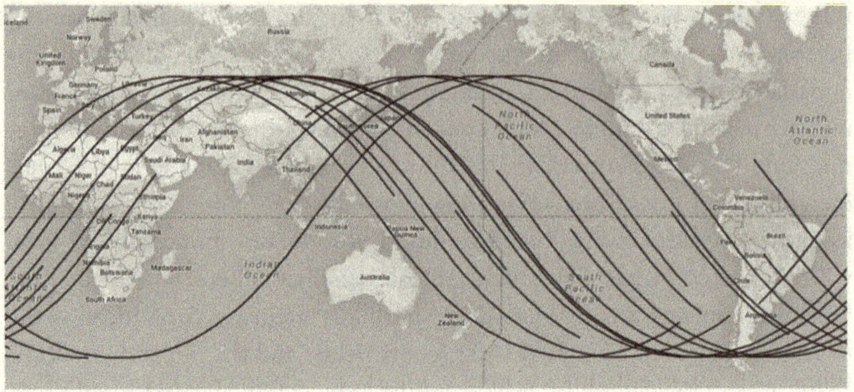

Figure 9.2 Orbital paths of an artifact "made in space," traversing multiple time zones.
Source: Made In Space.

Trifles and Tension

The extraordinary tensions at work that consolidate the fetish following Pietz and that are evident in the artifacts of attunement in LEO exemplify a number of qualities of the "extreme" identified by Battaglia et al. (2012) that characterize human habitation in space. As Battaglia and her colleagues suggest, the extreme conditions of outer space place the body and the artifact under novel tensions that produce new, emergent, and dynamic understandings of humanity "in the subjunctive mode." But the tensions that evidently emerge in the extreme conditions of LEO, as in other parts of outer space, I might hazard are one of degree, not of essential quality. These are matters of degree that I suggest are glaringly on view off Earth in LEO in the 21st century as well as historically off the coast of West Africa in the 16th and 17th centuries. These reveal what might be regarded as the fundamental "nuclear metaphysics" at play that the philosopher of object-oriented ontology, Harman (2009, 215–217), is at pains to explicate. In line with Harman and by focusing on this question of degree, I propose an inversion of conventional understandings regarding material culture, namely that such extreme conditions as those encountered in LEO as Pietz suggests apropos the revolutionary truth of the fetish assert a guarded and contingent "truth" apropos our understandings of material culture that hold both terrestrially and extraterrestrially.

Following Heidegger and contra Pietz, such tensions requiring "attunement" (following Gibson) or "comportment" (following Heidegger) are the general conditions of material life, despite the novel and extreme conditions off the coast of West Africa in the 16th and 17th centuries or the extreme conditions 400 km above the Earth in LEO. Such extreme conditions reveal certain contingent truths (or "justifications" to use Harman's term) that are repeatedly rehearsed regarding the nature of human material interactions that are qualitatively apparent and which belie received understandings of selfhood and materiality, such as the idiosyncratic Euro-American notion of the autonomous skin-enclosed self.

It is worthwhile here to consider Gorman's plastic bag, a "trifle" as Pietz would define it, in relation to the notion of the fetish but a "trifle" which by its very gloss embodies these elements of radical incommensurability as described by Pietz and one which attempts to impose a universal gravitational standard where there is not one naturally, much like Pietz's Akan gold weights.

These attunements serve to alleviate the inherent tensions present in the phenomenological apperception of the object, much as Harman's assertion regarding Heidegger's War Emergency Semester lectures following World War I, which Harman characterizes in dynamic indeterminate terms whose tensions were since glossed over in Heidegger's later and established static formulation of the fourfold and its subsequent variations. In discussing Heidegger's fourfold, Harman draws attention to the tensions inherent between the poles of the fourfold. Here, he observes, following Heidegger, that the relation between the four poles is "a sort of 'mirroring' relation'" (Harman 2009, 216). This is a "mirroring" that is also understood as a synonym in Heidegger's discussion of a "wedding" (Harman 2009, 216). As Carsten and Hugh-Jones (1995) comment in relation to Levi-Strauss' work on House Societies, the institution is consolidated and stabilized by "transfixing an unstable union, transcending the opposition between wife-givers and wife-takers and between descent and alliance, the house as an institution is an illusory objectification of the unstable alliance to which it lends solidity" (8). Harman (2009) writes further in reference to an improbably enduring solidity:

> We have seen that in the sensual realm, unified intentional objects endure even when viewed from countless different angles and distances and in ever-shifting moods. The intentional/sensual object is not the mythical "bundle of qualities," but a unit able to sustain an infinite number of "adumbrations." (217)

Key to Harman's observation is that these tensions can appear remarkably stable except when suddenly they are not: "The four tensions in the model each have their own way of breaking down under pressure and thereby triggering change" (Harman 2009, 220). It is precisely under the conditions of LEO where such seemingly stable tensions in fact do "break down" in relation to the Heideggerian fourfold, between the poles of the "earth and sky, divinities and mortals" (Heidegger 1993, 360). The gathering of these elements suggests that "there is no reason why this tension should not persist indefinitely, without seismic changes of any sort" (Harman 2009, 219), unless thrust into novel conditions (Harman 2009, 221)—other forms of gravity, for instance, be they micro, zero, and so on.

Despite the artifactual focus of this discussion, it is worthwhile to consider the body itself in its cyborgian aspect as just such a material artifact of attunement. But this is another discussion, where I would defer to my colleagues Jeevendrampillai and Parkhurst in our joint ETHNO-ISS project on the specific issues surrounding the body. However, the body itself might be considered to evoke the qualities of a "fetish" following Pietz, where the fetish is in fact an extension of the bodily nexus, producing an extreme territorialization and imbrication of the body reminiscent of Valentine's (2017b) and Battaglia et al.'s (2012) discussions of different forms of contingent and emergent dynamic forms of nature and cyborgian configurations in the "extremes" of space. Yet for the purposes of this discussion, I force an unnatural separation of elements within the cyborg body/artifact nexus (following Valentine's [2017b] lead). Under extraterrestrial conditions, such a distinction is untenable, there is no a priori bodily distinction between artifact and flesh, body, and environment—the cyborg is the natural condition (a new nature following Valentine [2017b]) of the inhuman and unnatural conditions of the human body according to terrestrial terms in space. Here, such terrestrial subjectifications are untenable, reflecting a long-standing theoretical repudiation of this Cartesian split but by no means foreign outside Euro-American contexts. In many respects, the cyborg explicates not a novel condition but possibly a more primordial condition that Harman (2009) intimates.

Novel Manufacturing and Novel Forms of Attunement

Most forms of off-world manufacturing to date highlight these novel and extraordinary scales of commensurability. Consider here fiber optics devised by Made In Space to be produced in LEO and then used on

Earth, whose superior uniform microscopic sections enable enhanced transmission unachievable under the conditions of Earth's gravity. Similarly, consider also the recent 3D printing on the ISS by Nscrypt (2020) of a human knee meniscus . Terrestrial conditions prevent the 3D printing of human tissue without a scaffold. In microgravity, such a scaffold is unnecessary, and a meniscus can be printed and more readily accepted by the human body, with the prospect of other body parts in the future. Such a meniscus introduces the possibility of the human body achieving a radically distinct "comportment" (Heidegger 2008b, 53) enabling new "contextures," to use Heidegger's language (55–60), of new (and "extreme" following Battaglia) configurations of radically and seemingly incommensurable space–times (Munn 1977, 1986), territorialities, and gravitational fields into a novel territorialization of the human body and its well-being literally of Earth and LEO (see also Valentine 2017b).

Such artifacts of attunement particularly in terms of their microscopic and seemingly "trifling" nature are at the heart of the perennial effects of the Copernican humiliation, the philosopher Blumenberg (1987) has addressed, just as Heidegger's horror of the image of the world from the Apollo spacecraft, destroyed "the world" more effectively than any atom bomb. Such shifts much like the celebrated Earthrise image which had to be rotated 90 degrees to alleviate this alienating and "humiliating" view to reassert the primacy of the world and Earth (Oliver 2015). Such trifling gestures echo Blumenberg's observations regarding small "trifling" actions as Archimedean gestures. Blumenberg (1987) describes how the figure of Copernicus is characterized as a "perpetrator" that inflicts "theoretical action as a violent deed" (262) which literally destroys "the world" at the center of a terracentric cosmology. Blumenberg comments in reference to Clacagnini's treatise on such a "violent deed" in Archimedean terms as small actions creating great effects, dependent on a trifling "little book," perpetrating the violent displacement of the Earth at the center of the universe "by the magical efficacy of the word" (Blumenberg 1987, 281)—echoing Harman's (2009) observations of this inherent instability.

Here, such "trifles" echo the sudden shifts of atmosphere in Stewart's (2011) discussion, where a slight shift of perception and its configuration produce terror and alienation where security and safety once prevailed where the processes of attunement that bring forth one world over another shift suddenly and radically. Apropos the conditions of microgravity in

LEO, such attunements or "comportments" (Heidegger) negotiate these tensions of incommensurability to produce a radical space–time (following Pietz) that is distinctly territorialized—not in terms of the "territory" of the 16th- and 17th-century fetish off the Muna Coast, but here territorialized in relation to LEO, profoundly upsetting conventional notions of territoriality (e.g., Made In Space's print run along the orbit of the ISS).

Rendering commensurate radically incommensurate elements can be observed in other contexts historically. Consider here the Byzantine understanding of the icon and its prototype as one of being a relative in the manner that the image of the father can be perceived in the image of the son. That is, the icon and its prototype is a relationship of kinship, despite its radically incommensurate elements of paint and wood and heavenly deities along its perceptual and material axis (the inherent incommensurability of which and its ultimately scandalous nature have always been the focus of iconoclastic scorn throughout history, which would insist on the "trifling" nature of icons). This is not unlike the relationships of kinship of affinity, between disparate elements, which is "marriage" literally and figuratively in Heideggerian terms by which the "tension" (following Harman) resolves the commensurability of these divergent elements. This in turn produces a new territorialization in terms of relations of affinity and co-residence not unlike the novel territorializations at the heart of Pietz's fetish. And this is not unlike what might be thought of in terms of the different forms of territorialization at the heart of the icon in Mondzain's discussion of the consolidation of Byzantine territorial sovereignty and the reversal of the icon's centrifugal motion (Mondzain 2004, 67) and the novel territory of the 3D printed artifact in LEO.

Incommensurable Affordances

However, this discussion is focused not on the affordances of 1 g or the novel affordances of 0 g (or microgravity) but instead on artifacts that attempt to reconcile 1 g with 0 g and vice versa and occupy both simultaneously—to in effect produce two sets of conflicting affordances simultaneously in hybrid conditions. Gibson and others note that affordances are multiple and situational in relation to a given phenomenon. They are adumbrated, subjunctive, and related but rarely, I argue, actually incommensurable, being universally understood as they are under the conditions of Earth's gravity.

Gibsonian affordance and Peircian indexicality share certain commonalities and exhibit certain divergences that are instructive here in terms of the novel configurations they enable. For example, fire within a Gibsonian framework "affords" burning with a delimited set of consequences (burnt flesh, pain, charred substance, etc., and of course smoke). This is not unlike Harman's (2009) observations concerning transductions from within the fourfold and the "adventures" of "heat." This is what Harman describes as "heat leaving the substance of fire and setting up shop in iron or water for a time [or] when the accidents of wine are transferred to the absent blood of Christ during the Eucharist" (220). In these terms, one might understand the affordance of a "clip" such as on Made In Space's wrench (Figure 9.3) is transduced, "sets up shop" so to speak following Harman, to produce the effects of 1 g in microgravity that hold it in place as a "gravity surrogate" (Gorman 2017).

Fire in a Peircean frame produces the indexicality of smoke. In some respects, "affordance" and "indexicality" are comparable and similar. Both produce a certain spacetime of the "affordances" and "environmental" context of fire and its consequences as well as the space–time of "indexicality." Both in turn suggest novel attenuated space–times. In both examples, this is directional, from the origin of the fire to the burn, the smoke (inwardly

Figure 9.3 A space-optimized, 3D printed wrench for ISS crew members.
Source: Made In Space.

[centripetally] and outwardly [centrifugally]) in two different and conflicting space–times which I argue produce innovative attunements because of this novel, shifting, and dynamic time–space. Similarly, not only are the affordance and indexicality of fire durational and directional (inhering in a fire source) but also they produce a new configuration of radically distanced elements that are in direct physical relation to one another (as noted by Harman 2009, 220). The index of smoke is empirically and physically linked to fire, just as burnt flesh is directly entailed in the fire itself. Here, one is reminded of Classical theories of haptic vision and their hazards (Crary 1992) where vision is a physical index of the source object, which transmit particles that impress themselves physically onto the viewer, as smoke indexes fire, as the burn scorches flesh, and as the affordance of fire and smoke enters the eyes; in both examples, the body is reconfigured into a distinctive time–space.

The innovation of the "icon" in early Christian traditions built upon the "habits" (Morgagni 2012) of earlier Antique theories of vision such as its haptic qualities to produce a new alignment of habits in terms of Christian piety and the novel universal Christian subject. This also produced a radically new alignment, and expansion and realms of experience in Christian cosmology that would link the pious earthly subject with the incommensurable realms of heaven. Through the "icon," one looks "past"—not "at"—the heavenly Christian realm and establishes a physical and spiritual contact which aligns earthly flesh with the heavenly through the communion of the radically reconfigured visual sphere of the icon and its expanding earthly territory that was at the heart of Byzantine imperial consolidation (Mondzain 2004). As the historian Patricia Cox Miller (2009) notes, early Christian pilgrims could look at the gaping wounds of desert ascetics where putrid "worms" were beheld as luminous "pearls" (Miller 1994, 146–147; 2009) indexing the domain of heaven and uniting the earthly and the heavenly. Two seemingly incompatible realms, Earth and heaven, are here aligned through the apprehension of "worms" of putrefaction as entailing a luminosity that unites two incompatible realms through this seemingly unsteady and decidedly tenuous qualisign, whose affordances produce the seemingly contradictory index of luminous worms/pearls, which a novel "comportment" (Heidegger) stabilizes in that instance of this particular manifestation of the Heideggerian fourfold. Here, contra Pietz, I hazard that the fetish not so much works in opposition to the idol or icon despite the insistence of Western observers but, rather, produces a distinctive territoriality and in particular a distinctive relation to and "comportment" of the sentient body that has learned to attend to the seemingly contradictory nature of these affordances: putrid worms as luminous pearls. And like the fetish following Pietz, the artifact of attunement is an emergent dynamic nexus of novel relations, which can only be hinted at here as they continue to emerge (or are "adumbrated" as Harman would have it).

The question of "attunement" in relation to incommensurate affordances relates to Stewart's 2011) use of the concept in her discussion of atmosphere. As Stewart writes,

> Incommensurate elements hang together in a scene that bodies labor to be in or to get through . . . bodies labor to literally fall into step with the pacing, the habits, the lines of attachment, the responsibilities shouldered, the sentience of a worlding. (452)

Such processes are at the heart of Heidegger's concept of "worlding," where a wide range of activities on the ISS serve to enable the attunement of terrestrial and extraterrestrial activities, such as the running of marathons on the ISS in time with those on Earth. This was something the British astronaut Tim Peake was able to do even more successfully, inhabiting co-temporally Greenwich Mean Time on the ISS, the same time zone as that in which the London Marathon took place. Similarly, Russian cosmonauts participate in "Immortal Regiment" celebrations simultaneously with those on Earth (see Buchli, 2025), in addition to numerous other religious observances and family events. But key to Stewart's discussion of attunement and of relevance here is the sudden way in which a given configuration changes radically that echo Harman (2009) in terms of transductions and their radical transformations, where heat from a fire is "setting up shop in Iron" as well as Blumenberg in his characterization of the Archimedean "perpetrator"—a given atmosphere switches from tranquility to dread in Stewart's discussion of shifting atmospheres and processes of attunement.

This dynamic and unstable quality which is the direct consequence of such incommensurable affordances might be profitably considered here in terms of the Heideggerian fourfold as a dynamic nexus of iconicity following Pierce. Any one individual element of the fourfold, that gathers together, works like the iconic nexus that gathers together disparate elements such as smoke and fire in a specific space–time (following Munn), seemingly disparate, nonetheless, they are in a novel and dynamic relation of intimacy. Heidegger's bridge at Heidelberg is a seemingly banal infrastructural intervention in the landscape—but a bridge—that gathers here the earth, and sky, heavens, and mortals into a distinct space–time and nexus of seemingly incommensurable elements that cannot be reduced one to the other, but exist within this dynamic nexus that constitutes Heidegger's notion of worlding and producing centripetal attunements and their affects and with them an interpenetration that constitutes the dynamic context of "being" described by Heidegger.

Harman's (2009) take on Heidegger's 1919 lecture in relation to the fourfold is characterized by a protean tension. Harman refers to this tension as a "nuclear metaphysics" related to the tensions inherent within any given object (216–217) as being also a mirroring. But this mirroring is also a "wedding," an observation Harman makes and attributes to Heidegger, who sometimes uses the term "wedding" as a synonym for "mirroring" (Harman 2009, 216). This suggests the play of elective affinities productive of a particular and novel space–time following the analogic thinking of Barbara Maria Stafford and the literal understanding of kinship and affinity—that is, as affines, relatives by marriage—the two irreconcilable elements reconciled within the deceptively stable institution of the "house" as described in Levi-Strauss' formulation (Carsten and Hugh-Jones 1995; see also Buchli 2013).

In this vein, the act of looking that characterized the working of the icon, that looked through the icon to the prototype, produced an intimacy and reconstitution of body and soul that was materially palpable. Vision here reconstitutes the body with all its contagious hazards, being a form of touch and thereby an attenuated worlding of body and heaven, and is here reprised in a surprising fashion under the conditions of LEO orbit, microgravity, 0 g, and other gravities, which can exist outside Earth's protective atmosphere, but where our ontological demands of vision and bodily apprehension are even more dangerous than they were for the societies of Mediterranean antiquity. The delicate engineering problem of the "window," on the cupola of the ISS in LEO and any other proposed extraterrestrial habitats (see Chapter 7, this volume) foregrounds the mortal dangers of vision and the levels of radioactive exposure that are at the heart of these visual economies which produce transfigurations of human consciousness, such as the "overview effect." Such attunements can be endured for only so long until radiation poisoning compromises human life. Such novel forms of "being" within these extraterrestrial conditions of the fourfold can be endured, territorialized, and made commensurable briefly until they come undone.

The examples of a weight, post, Ziploc bag, and wrench are presented here to demonstrate how radical incommensurabilities are made commensurable within what appear to be relatively banal artifacts. In particular, such artifacts of attunement that I propose here to exist in LEO are prime examples of what Harman might use to illustrate his concept of a "nuclear metaphysics" and the protean tensions entailed therein. As Harman (2018) observes the proverbial "hammer" in Heidegger's analysis that is "at hand" comes into analytical view when it is broken and falls apart, so too do artifacts of attunement in terms of their decidedly tenuous stability, accommodating incommensurable realms between 1 g and microgravity, that reveal the workings of such "nuclear metaphysics" that inhere in a given object and the divergent and incompatible realms they render contingently compatible and thereby render palpable how discordant realms are able to hang together and sustain human life while enabling insights into how such stable forms of life, elsewhere, like on Earth, are able to cohere.

Acknowledgments

This chapter emerged from a European Research Council Advanced Grant—ETHNO-ISS: An Ethnography of an Extra-terrestrial Society: The International Space Station. This project has received funding from the European Research Council under the European Union's Horizon 2020 research and innovation program (Grant No. 833135). The author is deeply indebted to the conversations and insights provided by the members of the team that have informed this chapter—Jo Aiken, Timothy Carroll, David Jeevendrampillai, Aaron Parkhurst, and Alica Okumura-Zimmerlinas—as well as to the editors of this book and their very thoughtful and insightful suggestions.

10
Space Virus 2020

Stefan Helmreich

Introduction

A few months into the COVID-19 pandemic, some friends invited my family over to their spacious backyard to look at Jupiter and Saturn. They had hooked their newly acquired telescope up to a camera, which fed images of the planets to a computer screen, in this way keeping each of our small family pods away from the instrument's eyepiece, which we worried could be a surface vector for the coronavirus. Because we were all social distancing, why not spend time with the insanely distant—or, better, join together in an attempt to bring the exceptionally distant safely close?

In those 2020 days, some of us found outer space—inaccessible, withdrawn—to be oddly comforting company. Next day, I received an email from a farther flung friend, who wrote,

> By the way, in case you are not already attuned, Jupiter and Saturn are in the Southern sky at sunset. And then around 8:30 or 9, Mars shows up, amazingly bright and red. In better days I would invite you over to look through my telescope. But for now, enjoy these companions, vast and cool and sympathetic.

It was like we were back in medieval times, when the sublunary sphere, the earthly realm below the Moon, was understood to suffer from corruptions from which the enduring cosmos was free.

The strange comfort provided by the uncontaminated otherworldly sits weirdly alongside occasional worries among astrobiologists about malevolent microbes from space. The field of astrobiology itself, known at its origins as exobiology, was in fact inaugurated with anxieties about extraterrestrial contamination coming to Earth. Nobel Laureate Joshua Lederberg, a bacterial geneticist at Stanford University, was in the early 1960s concerned about

Stefan Helmreich, *Space Virus 2020*. In: *Otherwhere Ethnography*. Edited by: Istvan Praet and Perig Pitrou, Oxford University Press. © Oxford University Press (2025). DOI: 10.1093/9780197790885.003.0011 the possibility that spacecraft returning to Earth might infect the planet with extraterrestrial microbes. His finger was on the popular pulse. In "Danger from Space," *Time* magazine in 1961 pronounced,

> The invaders most to be feared will not be little green Venusians riding in flying saucers or any of the other intelligent monsters imagined by science fictioneers. Less spectacular but more insidious, the invaders may be alien microorganisms riding unnoticed on homebound, earth-built spacecrafts. (Quoted in Wolfe 2002, 194)

Astrobiology began, then, as the biology of invasive alien species (the idea that biotic substance may have been ferried to Earth from space is still in circulation, and it was at the center of an orbital experiment led by Japanese astrobiologists from 2015 to 2018 [Yano et al. 2017]). Although Lederberg was also concerned with the contamination of other planets by Earth biota, public and funding focus centered on what was imagined as the "back contamination" of Earth by alien life. In "Germs in Space," historian Audra Wolfe (2002) notes that these concerns were saturated with Cold War imagery: "The American duty to protect freedom, through interplanetary settlement if necessary, might be challenged by invisible internal enemies" (195; and see Scoles 2020).

Fears of alien contamination fed the storyline of Michael Crichton's *The Andromeda Strain*, a 1969 science fiction novel (and later, movie) about an extraterrestrial microbe that arrives on Earth after hitching a ride on a military satellite, proceeding to tear a path of epidemic death beginning in Arizona and continuing on across the planet (and see Radin 2019). The microbe of *The Andromeda Strain* is alien in all kinds of ways, having no DNA or proteins, manifesting, rather, as a strange crystal form—although the military of the book/movie turns out to be prepared, not for its deadly specificity but, rather, for its fact; the biocontainment facility used to research and decode the alien organism happens to exist precisely because biological warfare is in the story already being researched by the U.S. government. *More*: The infected satellite itself, knocked off orbit by a meteorite, turns out to have been deliberately sent to search for upper atmosphere microbes that might be crafted into bioweapons. And it looks like the satellite indeed found such entities. Crichton's tale torques the ending of H. G. Wells's *War of the Worlds*, in which Earthly microbes, in the end, lay Martian invaders low.

In *The Andromeda Strain*, the microbial invaders are avatars of military-industrial-complex humans themselves. They have already been saved a place in an Earthly genealogy. Enemy organisms, hostile to Earthly vitality, they are *life forms* already anticipated—sought out, by the sociopolitical *forms of life* animating Cold War science and its anticipation of antagonism (on life forms and forms of life, see Helmreich 2009).

Move from Crichton's novel to the novel coronavirus, which has its own deep Earthly genealogy. In an August 2020 *Scientific American* opinion piece titled "The Ocean Carries 'Memories' of SARS-CoV-2," members of the Ocean Memory Project, a group of scientists and artists, refer readers to the field of paleoviromics, which holds that there exists "an ancient common ancestor for all RNA viruses, one that originated and diverged genetically in the ocean before jumping to land during the early evolution of terrestrial animals and plants." The ocean was early on a host to RNA viruses, and all organisms that have adapted to these entities have a kind of "memory" of them. The piece continues: "Mobile hosts like us, and the bat hosts before us, are the metaphorical seas and oceans in which viruses like SARS-CoV-2 persist." This vision of organisms as mobile seas has a thick lineage, embracing literature—consider Italo Calvino's (1969) "Blood, Sea," narrated by a volume of blood that recalls its primordial form—and science—consider geobiologist Mark and Dianna McMenamin's (1993) notion that the liquid in organisms on Earth forms a kind of worldwide web, what they call a "hypersea." Astrobiologists, too, have weighed in how microbial distributions in humans may have much in common with those in ocean ecosystems (Reid and Buckley 2011).

A common historical ecology for the emergence of humans and viruses, of course, does not make SARS-CoV-2 friendly—nor does its location in natural history extract it from *political* histories (which lab leak theories have made the center of their accounts), which now permit it to hitch a ride on, and to amplify, legacies of class, racial, national, age, and caste inequalities. It lives in what Mike Davis (1998) once called—speaking of Los Angeles, but the term can be generalized to thinking about the suffusing disasters of our day—an *ecology of fear*, one in which environmental threats (fire, flood, and pandemic) haunt the most everyday matters of living.[1]

Such ecologies, anchored in history, may nonetheless be experientially baffling, signs of frightening other worlds. Anthropologist Federico De Musso calls one of the spaces that emerged during the pandemic "Planet Corona," a realm that he named in June 2020 in his SF comic strip allegory about contracting COVID-19, along with his grandmother, which put them both on astronaut-like oxygen feeds before they recovered and came back (so many others did not) to this world. Worlds beyond Earth are here not the calming domain of the sublime that my friends invited me to see through their telescope but, rather, symbols of unease and alienation (Figure 10.1).

Contrast De Musso's moving allegory about space and corona with the May 2020 launch of SpaceX's *Crew Dragon* capsule, a public–private trip to the International Space Station undertaken during the late beginning of the COVID-19 pandemic. Anyone watching the split-screen coverage of White male COVID-cleared astronauts leaving Earth while the streets of Planet Corona saw protests for racial justice and reckoning in the wake of the police killings of George Floyd and Breonna Taylor—and of skyrocketing death rates from COVID for people of color—might be forgiven for flashing back to Gil Scott-Heron's 1970 composition, "Whitey on the Moon," which famously juxtaposed the trials of impoverished Black America simultaneous with the triumphalism of the Apollo landings:

> I can't pay no doctor bill. (but Whitey's on the moon)
> Ten years from now I'll be payin' still. (while Whitey's on the moon)[2]

Maybe the SpaceX moment was a sign that all of the early pandemic claims that we were "all in this together" were always dishonest. For all of the discourse about the global amity represented by the International Space Station, the celebration of getting off-Earth could not help but bring into relief that for many space boosters—think of the recent spate of billionaires in space—*leaving other people behind* is part and parcel of their dominant fantasies about space colonization (see Schultz's [2020] "Life as Exodus," an analysis of the escapist fantasies of Silicon Valley oligarchs). These depend on a grounding logic of what E. P. Thompson (1980) called the logic of *exterminism*, which "designates . . . characteristics of a society—expressed, in differing degrees, within its economy, its polity and its ideology—which thrust it in a direction whose outcome must be the extermination of multitudes" (22). Such a logic may be more or less implicit—although SpaceX CEO Elon Musk's demand, in May 2020, that his Tesla car manufacturing facilities be exempt from California's state and local shelter-in-place orders, meant to slow the spread of COVID-19, fairly explicitly indicates his willingness to put other people's lives at risk for his own projects of technological acceleration.

Figure 10.1 Panels from Federico De Musso's "Voyage to Corona," 2020.

What of anthropologies of space/off-Earth? Following Lisa Messeri and Valerie Olson (Chapter 1, this volume), it is worth asking how scholars working in this domain might address the sharp possibility that anthropologies of space themselves may reproduce some of the same exclusions of many space programs, particularly if, as with U.S. American cases, they do not grapple with (1) the settler colonial fantasies of frontiering that subtend the history of space programs and aspirations (cue Jeffrey Bezos' idiot cowboy-hatted jaunt to near-Earth space in 2021) (and see Shorter and TallBear 2021) and (2) the racialized politics—*and, specifically, the Whiteness*—of interstellar colonial fantasy. In "The Case for Letting Anthropology Burn," Ryan Cecil Jobson (2020) reminds us that anthropology, particularly in the United States, in the wake of Boasian critiques of biological determinism—which displaced racial antagonisms from the violence of genocide and plantation slavery onto the critique of biological determinism (important in its own way, but not where race and racism live in everyday life)—has made much of anthropology into a "'white public space' that maintains a liberal myth of perfectibility through the progressive incorporation of historically subordinated peoples into the comforts and privileges of property and citizenship" (265). One might worry about anthropologies of outer space operating through similar logics of "incorporation"—attentions to, for example, Indigenous science fiction stories and movies in ways that do not take aim at the political economies that make space programs increasingly tuned to private capital or that treat decolonization as a metaphor (see Tuck and Yang 2012) rather than material practice about land and space.

Images of imperial contest in space, strangely, leaked into early national competitions to develop a vaccine. The United States and Russia, in their race for ready inoculation, reanimated Space Age imagery and rivalries. Russia's vaccine was early on called the *Sputnik*, after the 1957 satellite that beat the United States into orbit. The imagery on the dedicated website was astonishing, with a video depicting Earth as encased in the coronavirus, with an orbiting *Sputnik V* swooping circles around the planet as the craft/vaccine pokes holes in the viral sphere in order to free the Earth of the pathogen's embrace (Figure 10.2).

The United States, meanwhile, with its "Operation Warp Speed" project—its name a show-biz call back to the *Star Trek* science-fiction television show—offered a logo meant to look like a military patch, featuring an image of the coronavirus against a backdrop jumble of symbols: the abstract bird of the U.S. Department of Health and Human Services logo (whose winged feathers are visible peeking out on the right), five stars representing the five government groups partnering in this project, and a mysterious grounding hexagon that reminds me of the hexagonal microbes at the center of the Andromeda Strain (Figure 10.3).[3]

Strange that these projects, aimed at a subvisible virus, should in their rhetoric balloon up to outer-space proportions. There was a kind of "derangement of scale" (T. Clark 2012) at work here. Such a view-from-outer-space belied, I suggest, the early failure of many public health bodies to attend to the spatialities (*not* outer spaces) that have actually shaped COVID, from de facto racially segregated neighborhoods to isolated

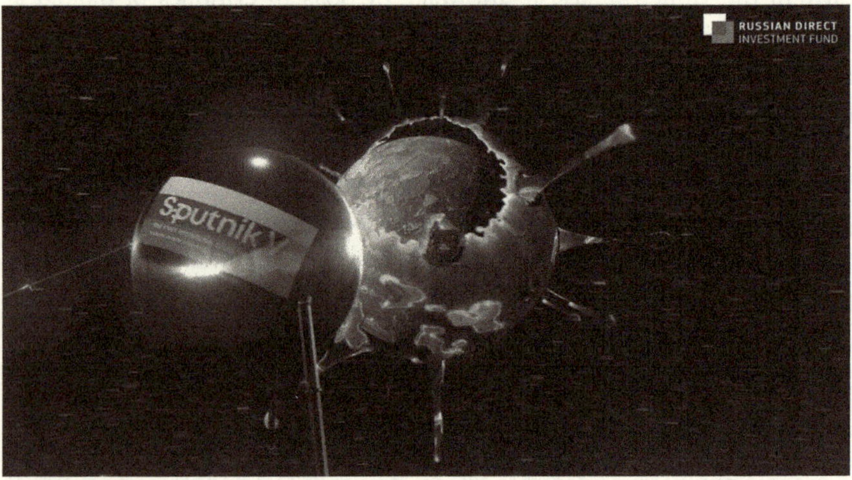

Figure 10.2 Publicity image for Russia's Sputnik vaccine.

Figure 10.3 Logo for the United States' Operation Warp Speed.

and underprepared nursing homes, to prisons, to meat packing facilities. These are not spatialities to be grappled with from some spacey aerie.

In the event, vaccine battles ended up fought less in the zone of science moonshots than in the more sublunary realm of anti-vax politics and of nationalist and pharmaceutical protectionisms (Banerjee 2021). And by March 2022, when I was revising this chapter—first drafted in October 2020—Putin's Russia had opted not for a path of healing people but, with its attack on Ukraine, for a plan of killing them, a path not aimed at any kind of planetary protection, but at a kind of 19th-century fantasy of empire on Earth. With updated 21st-century nuclear weapons added to that mix, concerns about Cold War air power reactivated, while concerns about COVID air profiles fell into the background.

Notes

Chapter 2

1. As Danielle Briot has noted, "As early as 1914, on the basis of observations of Earthshine, he concluded that Earth seen from space has to be seen with a pale blue color." Danielle Briot. 2013. "The Creator of Astrobotany, Gavriil Adrianovich Tikhov," in *Astrobiology, History, and Society: Advances in Astrobiology and Biogeophysics*, edited by D. Vakoch. Berlin: Springer-Verlag, p. 175. The secondary literature on Tikhov is quite scarce, but see also Danielle Briot, Jean Schneider, and Luc Arnold. 2004. "G. A. Tikhov and the Beginnings of Astrobiology." In *Extrasolar Planets: Today and Tomorrow: ASP Conference Series* 321, edited by Jena-Philippe Beaulieu, Alain Lecavelier des Etangs, and Caroline Terquem, pp. 219–220. San Francisco: Astronomical Society of the Pacific; Victor Tejfel. 2010. "Gavriil Adrianovich Tikhov (1875–1960): A Pioneer in Astrobiology." *Highlights in Astronomy* 15: 720–721; and Tuken B. Omarov and Bulat T. Tashenov. 2005. "Tikhov's Astrobotany as a Prelude to Modern Astrobiology." In *Perspectives in Astrobiology*, vol. 366, edited by Richard B. Hoover, Alexei Yu. Rozanov, and Roland Paepe, pp. 86–87. NATO science series, series 1: Life and Behavioural Sciences. Brussels: NATO.
2. See Luis A. Campos. 2020. "Life as It Could Be." In *Social and Conceptual Issues in Astrobiology*, edited by Kelly Smith and Carlos Mariscal, pp. 101–114. New York: Oxford University Press; and Luis A. Campos. 2019. "Genetic Sputniks." *Grow Magazine.*
3. Gavriil Tikhov. 1959. *Sixty Years at the Telescope*. Moscow: Detgiz.
4. Gavriil Tikhov. 1962. *L'enigme des planètes*. Sablons, France: En Langues Etrangeres.
5. "Je ne pouvais alors prévoir que plusieurs années plus tard la botanique et l'astronomie se rejoindraient dans ma pensée et qu'aux confins de ces deux disciplines apparaîtrait l'astrobotanique, une science réunissant les lointaines planètes et la vie des plantes" (p. 29).
6. Tikhov, *Sixty Years at the Telescope.*
7. Tikhov, *Sixty Years at the Telescope*, p. 85.
8. Tikhov, *Sixty Years at the Telescope.*
9. Tikhov, *Sixty Years at the Telescope*, pp. 69–70.
10. Tikhov, *Sixty Years at the Telescope.*
11. Tikhov, *L'enigme des planètes*, 1962, p. 95.
12. Tikhov, *Sixty Years at the Telescope*, p. 97.
13. "Et j'eus le bonheur insigne d'être le premier dans l'histoire de la science à prononcer à haute voix le mot d'astrobotanique" (p. 110). Hubertus Strughold would later mistakenly attribute the first appearance of the term "astrobotany" to a pamphlet published by Tikhov in 1947. Hubertus Strughold. 1959. "Advances in Astrobiology." In *Proceedings of the Lunar and Planetary Exploration Colloquium*, 1, no. 6, April 25, pp. 1–7.
14. "Philosopher and optician Giambattista della Porta would be bemused. In 1589 he proposed, in his *Magia Naturalis*, projecting earthbound life onto the surface of the moon through a parabolic mirror, using the moon as a screen." Brunner, Bernd. 2011. *Moon: A Brief History*. New Haven, CT: Yale University Press.
15. Gavriil Tikhov, 1960. *Principal Works: Astrobotany and Astrophysics, 1912–1957*. Washington, DC: U.S. Joint Publications Research Service. p. 83.
16. Tikhov, *Sixty Years at the Telescope.*
17. Tikhov, *Principal Works*, pp. 38, 44. As Strughold would later explain Tikhov's views, "He reported that the colder the climate, the less is the reflecting power of plants in the main heat-carrying rays from infrared to red and yellow. Optically, this means their color is shifted to the bluish side. Ecologically, it means that they absorb more heat. Since the dark areas on Mars show a strong bluish green tint, Tikhov thinks that the plants on Mars have developed just these optical properties as an adaptation to the severe Martian climate. He generalizes this theory and applies it to the other planets, stating that the optical properties manifested in the

color of plants essentially reflect adaptation to the general level of the environmental temperature. On Mars, therefore, where the climate is vigorous, the plants are of blue shades; on Earth, where the climate is intermediate, the plants are green; and on Venus, where the climate is hot, the plants have orange colors, says Tikhov." Strughold, "Advances in Astrobiology," p. 3.

18. Luis A. Campos. 2017. "Dialectics Denied: Muller, Lysenkoism, and the Fate of Chromosomal Mutation." In *The Lysenko Controversy as a Global Phenomenon: Genetics and Agriculture in the Soviet Union and Beyond*, edited by William de Jong-Lambert and Nikolai Krementsov, pp. 161–184. New York: Palgrave Macmillan.
19. Tikhov, *Sixty Years at the Telescope.*
20. Tikhov, *Principal Works*, p. 34.
21. Tikhov, *Sixty Years at the Telescope.*
22. Tikhov, *Principal Works*, pp. 59, 63. "(The Sector of Astrobotany turned to the Presidium of the Academy of Sciences Kazakh SSR with the request that a discussion be organized on the problem of extra-terrestrial life, which was then conducted on 25–27 September 1952 at the city of Alma-Ata. * Footnote: 'At the time of the Alma-Ata discussion, astrobotany had already indicated the ways of studying microorganic life on the giant planets, and it had begun to be called astrobiology.')" According to Hubertus Strughold, "Astrobiology, as a book title, appeared first in a publication by Gabriel Tikhov in 1953. I used the word in the same year in the text of *The Green and Red Planet*, and Flavio A. Pereira, of Brazil, wrote in 1956 a book entitled *Introduction to Astrobiology*. Hubertus Strughold, "Advances in Astrobiology".
23. Tikhov, *Principal Works*, p. 63.
24. Alan Dunn. 1960. *Is There Intelligent Life on Earth? A Report to the Congress of Mars, Translated into English by the Author.* New York: Simon & Schuster.
25. Tikhov, *Principal Works*, p. 34.
26. Tikhov, *Principal Works*, p. 84.
27. "Nowadays, in order to prepare for the detection of life in remote extrasolar planets, astronomers observe Earthshine to detect the spectrum of terrestrial chlorophyll, and specifically the Vegetation Red Edge (VRE) in the near infrared" (Briot, "The Creator of Astrobotany"). See also Istvan Praet. 2017. "Astrobiology and the Ultraviolet World." *Environmental Humanities* 9(2): 378–397; and Nancy Y. Kiang. 2008. "The Color of Plants on Other Worlds." *Scientific American* 298(4): 48–55.
28. Western group on Planetary Biology, February 21, 1959. Folder "OR NAS, Space Science Board, Committees: WESTEX, Ad hoc," Space Science Board Papers, National Academies of Science; see also Box 981, Carl Sagan Papers, Library of Congress.
29. Martin Freundlich and Bernard E. Wagner, eds. 1967. *Exobiology: The Search for Extraterrestrial Life, Volume 19, American Astronautical Society, Science and Technology Series.* AAS/AAAS Symposium, New York, December 30.
30. I. Ivanov, pseudonym of Andrei Remizov. G. A. Tikhov. 1955 "Is Life Possible on Other Planets?" *Journal of the British Astronomical Association* 65(3): 193–204.
31. March 10, 1959, Folder "OR NAS, Space Science Board, Committees: WESTEX, Ad hoc," Space Science Board Papers, National Academies of Science. As Steven J. Dick has noted, "In 1947 at the famous University of Chicago conference on planetary atmospheres Kuiper (1949, *Atmospheres of the Earth and Planets*) postulated plants similar to lichens on Mars, and in 1957 and 1959 Sinton gave his widely accepted spectroscopic proof of vegetation on Mars." Steven J. Dick. 1991. "From the Physical World to the Biological Universe: Historical Developments Underlying SETI." In *Bioastronomy. Lecture Notes in Physics*, vol. 390, edited by J. Heidmann and M. J. Klein. Berlin: Springer, pp. 356–363, on 360.

Chapter 3

1. "Jupiter's Moon Io: A Flashback to Earth's Volcanic Past," NASA press release by Douglas Isbell (NASA headquarters, Washington, DC) and Jane Platte (Jet Propulsion Laboratory, Pasadena, CA), November 19, 1999, available at https://www.jpl.nasa.gov/news/jupiters-moon-io-a-flashback-to-earths-volcanic-past.
2. Tracy Gregg, interviewed by Ellen Goldbaum, "From Lava Lakes on Jupiter's Moon, Io, Come Ideas About What Earth May Have Looked like as a Newborn Planet," March 19, 2004, available at https://www.buffalo.edu/news/releases/2004/03/6629.html.
3. Gregg, interviewed by Goldbaum, "From Lava Lakes on Jupiter's Moon."

4. Fraser Cain, "Does Io Look like an Early Earth?" *Universe Today*, March 22, 2004, available at https://www.universetoday.com/9419/does-io-look-like-an-early-earth.
5. Paul Scott Anderson, "More Evidence That Europa's Ocean Is Habitable," *EarthSky*, June 28, 2020, available at https://earthsky.org/space/europa-ocean-habitable-goldschmidt-conference; See also M. Melwani Deswani and S. D. Vance, "Evolution of Volatiles from Europa's Interior into Its Ocean," 2020, available at https://goldschmidtabstracts.info/abstracts/abstractView?doi=10.46427/gold2020.1777.
6. Melwani Deswani and Vance, "Evolution of Volatiles."
7. John Raphael, "NASA: Jupiter's Moon Europa Is Surprisingly More Earth-like Than Previously Thought," *Nature World News*, May 18, 2016, available at https://www.natureworldnews.com/articles/22466/20160518/nasa-jupiters-moon-europa-surprisingly-more-earth-previously-thought.htm.
8. David Grinspoon, "The Earth in Human Hands," March 2017, interview available on YouTube at https://www.youtube.com/watch?v=1wD_cYEixYM. The relevant passage is in the 52nd minute.
9. See, e.g., Alan Boss, "From Pale Red Dot: 'Pale Blue Dot, Pale Red Dot, Pale Green Dot . . .,'" *Carnegie Institution for Science*, January 14, 2016, available at https://sciencesprings.wordpress.com/2016/01/20/from-pale-red-dot-pale-blue-dot-pale-red-dot-pale-green-dot.
10. Today, it is generally accepted that modern phyla such as the Bilateria and the Cnidaria originated in the Ediacaran period, but that finding is irrelevant for our purposes here. A few Ediacaran fossils have been interpreted as stem group metazoans, but the Cambrian period marks the unequivocal appearance of most major phyla (Droser and Gehling 2015, 4865).

Chapter 5

1. This work is based on the research supported by the South African Research Chairs Initiative of the Department of Science and Technology and National Research Foundation of South Africa (Grant No. 98765).
2. For more on these projects, see www.cosmopolitankaroo.co.za.
3. See, e.g., https://slate.com/technology/2017/03/why-we-need-to-stop-talking-about-space-as-a-frontier.html.
4. Robert Braun, SKA director of science, https://www.skatelescope.org/newsandmedia/outreachandeducation/skawow/transformational-science.
5. As of February 2024, the billboard has been removed.
6. The Karoo is the name given to two sparsely populated biomes in the western half of South Africa, much of it uncultivated rangeland given over to extensive sheep farming, that constitute approximately 30% of the country.
7. MeerKAT was developed by SKA South Africa (SKA SA), later subsumed within South African Astronomical Observatory (SARAO) and funded by the South African Department of Science and Technology (DST).
8. The installation of the first SKA dish was announced on April 4, 2024. See, e.g., https://mybroadband.co.za/news/science/531131-the-big-lift-first-dish-of-worlds-biggest-radio-telescope-in-south-africa-going-up-soon.html.
9. The composition of the SKA organization is still evolving. Whereas Germany withdrew from the consortium, becoming an observer country in 2014, France and Spain joined it in 2018.
10. Interview with the SKA Vice Director, Manchester, June 27, 2016.
11. https://www.skatelescope.org/news/dual-site-agreed-square-kilometre-array-telescope.
12. Interview with the SKA Director, Manchester, June 28, 2016.
13. https://www.bbc.com/news/science-environment-18194984.
14. https://www.sarao.ac.za/land-rezoning.
15. https://www.saeon.ac.za/enewsletter/archives/2017/december2017/doc02.
16. Ibid.

Chapter 7

1. For now at least. NASA's Earth Observatory has been monitoring the annual depletion of ozone in the stratosphere above Antarctica for decades, and although industrialization had taken its toll, the banning of chlorofluorocarbons and other practices have led to variable years of repair.

Chapter 8

1. The ethnographic data on which I am drawing in this chapter were collected during an Astrobiology Summer School titled "Biosignatures and the Search for Life on Mars," which took place in Iceland in July 2016. The summer school was co-organized by the European Astrobiology Campus, the Nordic Network of Astrobiology (institutions from Sweden, Denmark, Finland, Norway, Iceland, Estonia, Lithuania, and the United States participate), and the COST Action "Origins and Evolution of Life in the Universe." We totaled 39 students, at very different career stages—from master's students to postdocs—and from different countries—13 from the United States, 24 from Europe, 1 from China, and 1 from Brazil. The first week of lessons was followed by group activities in the field.
2. An interesting discussion of the ambiguities embedded in the term exploration in the space science context can be found in Clancey (2012, 42 ff). See also Messeri (2016, 18).
3. Planetary scientists often use the expression "Earth-like" as well; interestingly, however, although many places on Earth are somehow said to be Mars-like or Moon-like, no place in the solar system is said to be Earth-like. In the context of exoplanet research, on the other hand, the search for an Earth-like planet orbiting around another star is now in full swing, but the term Earth-like refers to the size of the planet and its distance from the star; no reference is possible (so far) to ground environmental conditions.
4. For an interesting discussion on the epistemological dependence of field and laboratory, see Kohler (2002).
5. They call this *processual* approach to knowledge *local traditional knowledge* (LTK): LTK is not inside people's head, but it is out there in the environment and it consists in the skillful engagement with it; as such, keeping LTN alive requires a dose of improvisation, like in a jazz performance. LTK is contrasted to *modern traditional knowledge* (MTK), the approach to knowledge that understands tradition as a *substance*, something that can be passed on from a generation to another, ideally without mutations, changes, or adaptations.
6. At the end of the article by Ingold and Kurttila (2000), the authors claim that scientists' field activities are a form of LTK. Elsewhere, nevertheless, Ingold mentions the Mars rovers as an example of a dull attempt to find life without any engagement with the environment. In line with what is argued in this chapter, I believe that the Martian rovers operated by NASA engineers and planetary scientists are interesting cases to consider sociologically and can indeed be considered a form of LTK. For an interesting example of driving a rover as embodied practice, see Vertesi (2015).
7. Ingold refers to the Koyukon people's practices of naming as an example of storied knowledge, opposed to the classificatory vocation of scientific practices of knowledge making, and yet, he admits that the picture of scientific knowledge as antithetical to inhabitants' knowledge is somewhat idealized. In one of his many formulations, he writes that "contrary to the official view, what goes for inhabitant knowledge also goes for science. In both cases, knowledge is integrated not through fitting local particulars into global abstractions, but in the movement from place to place, in wayfaring. Scientific practices have the same place-binding (but not place-bound) character as the practices of inhabitants. Science, too, is meshworked" (2011, 154).
8. In fact, this aspect has always been subsumed in sociological finitism: Ostension (the act of showing by pointing, which lies at the very heart of meaning finitism as formulated by Bloor, Barnes, and Henry) is a remarkable example of the role of the body in the learning process. For a different but equally interesting account of pointing as situated practice, see Goodwin (1990).
9. Certain sites are used as analog for present Mars—for example, when trying instruments and sampling procedures; other sites are considered as analogs for past Mars, when volcanoes and glaciers interacted. In the second case, water is not a difference but, rather, a fundamental similarity. These two uses of Iceland as a Mars analog coexist (as we will see later), making water both important and invisible.
10. Ignorance is as social as knowledge. See Gross (2007), Marcheselli (2020), McGoey (2012) and Smithson (2007).
11. This included a few pairs of sterile gloves, three or four masks, an equal amount of sterile spatulas, and 14 falcon tubes; we were also allowed to use what we had with us—for example, a geologist's hammer or a Swiss army knife—and what we were able to find in the local shops, such as aluminum foil and alcohol.

12. We had been warned that this experiment would not work if searching for alien life: ATP, the molecule that mediates in the metabolic processes of storage and consumption of energy, is indispensable for Earthly life, but it is very difficult to imagine that the same sophisticated mechanisms might have evolved independently elsewhere.
13. The huge eruption in Holuhraun lasted for 181 days. It began on August 31, 2014, and ended on February 27, 2015. The new lava field covers 85 km^2.
14. Private conversation, October 19, 2015.
15. Most of the time, I suspect, the actors involved in space exploration in the present rue the fact that, most likely, they will never set foot on another planet; thus, they enjoy the feeling by proxy.

Chapter 10

1. Davis famously warned of a global pandemic in his 2005 *The Monster at Our Door*.
2. And see Lewis (2012).
3. In October 2020, I contacted the office responsible for Health and Human Services Logo, Seal and Symbol Policies, and they were not able to tell me the meaning of the hexagon. The officer who returned my email said that they had contacted the Biomedical Advanced Research and Development Authority, but that this office did not know either.

References

Ackmann, Martha. 2003. *The Mercury 13: The True Story of Thirteen Women and the Dream of Space Flight.* Random House.

Adams, Douglas. 1983. *The Restaurant at the End of the Universe.* Pocket Books.

African Union Commission. 2014. "Science, Technology and Innovation Strategy for Africa 2024 (STISA-2024)." https://au.int/en/documents/20200625/science-technology-and-innovation-strategy-africa-2024

Aït-Touati, Frédérique, Alexandra Arènes, and Axelle Grégoire. 2022. *Terra Forma: A Book of Speculative Maps.* MIT Press.

Anderson, Ryan, Emma Louise Backe, Taylor Nelms, Elizabeth Reddy, and Jeremy Trombley. 2018. "Speculative Anthropologies." *Fieldsights*, December 18. https://culanth.org/fieldsights/series/speculative-anthropologies

Appadurai, Arjun. 2013. *The Future as Cultural Fact: Essays on the Global Condition.* Verso.

Apple, Russell A. 1987. "Thomas A. Jaggar, Jr., and the Hawaiian Volcano Observatory." In *Volcanism in Hawaii: U.S. Geological Survey Professional Paper 1350*, edited by Robert W. Decker, Thomas L. Wright, and Peter H. Stauffer, Vol. 2. U.S. Geological Survey.

Arney, Giada, Shawn D. Domagal-Goldman, Victoria S. Meadows, et al. 2016. "The Pale Orange Dot: The Spectrum and Habitability of Hazy Archean Earth." *Astrobiology* 16(11): 873–899.

Atalay, Sonya, William Lempert, David Delgado Shorter, and Kim TallBear. 2021. "Indigenous Studies Working Group Statement." *American Indian Culture and Research Journal* 45(1): 9–18.

Baba, Marietta. 1999. "The Globally Distributed Team: Learning to Work in a New Way, for Corporations and Anthropologists Alike." *Practicing Anthropology* 23(4): 2–8.

Bagnold, Ralph Alger. 1941. *The Physics of Blown Sand and Desert Dunes.* Methuen.

Banerjee, Dwaipayan. 2021. "From Internationalism to Nationalism: A New Vaccine Apartheid." *Comparative Studies of South Asia, Africa and the Middle East* 41(3): 312–317.

Barge, Laura M., and Lauren M. White. 2017. "Experimentally Testing Hydrothermal Vent Origin of Life on Enceladus and Other Icy/Ocean Worlds." *Astrobiology* 17(9): 820–833. https://doi.org/10.1089/ast.2016.1633

Barnes, Barry, David Bloor, and John Henry. 1996. *Scientific Knowledge: A Sociological Analysis.* Athlone.

Barrera-Osorio, Antonio. 2006. *Experiencing Nature: The Spanish American Empire and the Early Scientific Revolution.* University of Texas Press.

Battaglia, Debbora, ed. 2006. *E.T. Culture: Anthropology in Outer Spaces.* Duke University Press.

Battaglia, Debbora. 2012. "Coming in at an Unusual Angle: Exo-Surprise and the Fieldworking Cosmonaut." *Anthropological Quarterly* 85(4): 1089–1106.

Battaglia, Debbora, Valerie Olson, and David Valentine. 2012. "Introduction: Extreme Limits and Horizons in the Once and Future Cosmos." *Anthropological Quarterly* 85(4): 1007–1026.

Battaglia, Steven M., Michael A. Stewart, and Susan W. Kieffer. 2014. "Io's Theothermal (Sulfur)–Lithosphere Cycle Inferred from Sulfur Solubility Modeling of Pele's Magma Supply." *Icarus* 235: 123–129.

Beaton, Kara H., Steve P. Chappell, Andrew F. J. Abercromby, et al. 2017. "Extravehicular Activity Operations Concepts Under Communication Latency and Bandwidth Constraints." In *2017 IEEE Aerospace Conference*. IEEE.

Beaton, Kara H., Steven P. Chappell, Andrew F. J. Abercromby, et al. 2019. "Using Science-Driven Analog Research to Investigate Extravehicular Activity Science Operations Concepts and Capabilities for Human Planetary Exploration." *Astrobiology* 19(3):300–320.

Becker, Joffrey. 2018. "Les Forces de l'attraction. Comment Ramener la Lune sur Terre." http://www.joffreybecker.fr/pdf/JBecker_LesForcesDeLAttraction.pdf

Becker, Joffrey. n.d. "Meeting the Alien: Experimenting with Post-Biological Forms of Life." Paper presented at Ethnographies of Outer Space: Methodological Opportunities and Experiments, University of Trento, Department of Sociology and Social Research, September 2022, Trento, Italy.

Behar, Katherine, ed. 2016. *Object-Oriented Feminism*. University of Minnesota Press.

Bell, Katherine L. C., Chris German, Zara Mirmalek, and Amy Pallant. 2015. "Transforming Remotely Conducted Research Through Ethnography, Education & Rapidly Evolving Technologies (TREET)." *Oceanography* 28(1): 40–43.

Benjamin, Ruha. 2016. "Racial Fictions, Biological Facts: Expanding the Sociological Imagination Through Speculative Methods." *Catalyst: Feminism, Theory, Technoscience*, 2(2): 1–28.

Bennett, Jane. 2010. *Vibrant Matter: A Political Ecology of Things*. Duke University Press.

Berry, W. 1977. "Mr. Gerard O'Neill's Space Colony Project." In *Space Colonies: A Coevolution Book*, edited by Stewart Brand. Penguin.

Bezos, Jeff. 2019. "Going to Space to Benefit Earth. Blue Origin." YouTube, May 10. https://www.youtube.com/watch?v=GQ98hGUe6FM

Bijker, Wiebe E., and John Law. 1992. *Shaping Technology/Building Society: Studies in Sociotechnical Change*. Cambridge, MA: MIT Press.

Black, Alexis D. 2018. "Wor(l)d-Building: Simulation and Metaphor at the Mars Desert Research Station." *Journal of Linguistic Anthropology*, *28*(2): 137–155.

Blumenberg, Hans. 1987. *The Genesis of the Copernican World*. MIT Press.

Boehrer, Bruce. 2010. *Parrot Culture: Our 2500-Year Fascination with the World's Most Talkative Bird*. University of Pennsylvania Press.

Bond, Alan B., and Judy Diamond. 2019. *Thinking like a Parrot: Perspectives from the Wild*. Chicago University Press.

Bowker, Geoffrey C., and S. Leigh Star. 1999. *Sorting Things Out*. MIT Press.

Braidotti, Rosa. 2006. "Posthuman, All Too Human: Towards a New Process Ontology." *Theory, Culture & Society* 23(7–8): 197–208.

Bratton, Benjamin. 2019. *The Terraforming.* Strelka Press.

Briot, Danielle. 2013. "The Creator of Astrobotany, Gavriil Adrianovich Tikhov." In *Astrobiology, History, and Society: Advances in Astrobiology and Biogeophysics*, edited by Douglas A. Vakoch. Springer.

Briot, Danielle, Jean Schneider, and Luc Arnold. 2004. "G. A. Tikhov and the Beginnings of Astrobiology." In *Extrasolar Planets: Today and Tomorrow: ASP Conference Series 321*, edited by Jena-Philippe Beaulieu, Alain Lecavelier des Etangs, and Caroline Terquem. Astronomical Society of the Pacific.

Brown, Alison. 2010. *The Return of Lucretius to Renaissance Florence.* Harvard University Press.

Brunner, Bernd. 2011. *Moon: A Brief History.* Yale University Press.

Buchli, Victor. 2013. *An Anthropology of Architecture.* Bloomsbury.

Buchli, Victor. 2020. "Extraterrestrial Methods: Towards an Ethnography of the ISS." In *Lineages and Advancements in Material Culture Studies, Perspectives from UCL Anthropology*, edited by A. Walford, T. Carroll, and S. Walton. Routledge.

Buchli, Victor. 2025. "A Speculative Ethnographers Guide to Low Earth Orbit." In *Anti-Atlas: Towards a Critical Area Studies*, edited by W. Bracewell. UCL Press.

Byrd, Jodi A. 2011. *The Transit of Empire: Indigenous Critiques of Colonialism.* University of Minnesota Press.

Cabrol, Nathalie. 2018. "The Coevolution of Life and Environment on Mars: An Ecosystem Perspective on the Robotic Exploration of Biosignatures." *Astrobiology* 18(1): 1–27.

Cabrol, Nathalie. 2022. "Voyage au Frontières de la vie." YouTube. https://www.youtube.com/watch?v=NmF9bUPY7s4

Calvino, Italo. 1969. "Blood, Sea." In *T Zero.* Harcourt, Brace & World.

Campos, Luis A. 2017. "Dialectics Denied: Muller, Lysenkoism, and the Fate of Chromosomal Mutation." In *The Lysenko Controversy as a Global Phenomenon: Genetics and Agriculture in the Soviet Union and Beyond*, edited by William de Jong-lambert and Nikolai Krementsov. Palgrave Macmillan.

Campos, Luis A. 2019. "Genetic Sputniks," *Grow Magazine.* Ginkgo Bioworks.

Campos, Luis A. 2020. "Life as It Could Be." In *Social and Conceptual Issues in Astrobiology*, edited by Kelly Smith and Carlos Mariscal. Oxford University Press.

Campos, Luis. n.d. "*Homo universalis: Queering* Habitability." Paper presented at Ethnographies of Outer Space: Methodological Opportunities and Experiments, University of Trento, Department of Sociology and Social Research, September 2022, Trento, Italy.

Canfield, Donald E. 2015. *Oxygen: A Four Billion Year History.* Princeton University Press.

Capova, Klara A. 2013. "The Detection of Extraterrestrial Life: Are We Ready?" In *Astrobiology, History, and Society*, edited by D. Vakoch. Springer.

Carr, M. H. 1996. *Water on Mars.* Oxford University Press.

Carsten, Janet, and Stephen Hugh-Jones, eds. 1995. *About the House: Lévi-Strauss and Beyond.* Cambridge University Press.

Castaño, Paola, 2021. "From Value to Valuation: Pragmatist and Hermeneutic Orientations for Assessing Science on the International Space Station." *The American Sociologist* 52: 671–701.

Casumbal-Salazar, Iokepa. 2017. "A Fictive Kinship: Making 'Modernity,' 'Ancient Hawaiians,' and the Telescopes on Mauna Kea." *Native American and Indigenous Studies* 4(2): 1–30.

Cavalazzi, Barbara, R. Barbieri, F. Gómez, et al. 2019. "The Dallol Geothermal Area, Northern Afar (Ethiopia)—An Exceptional Planetary Field Analog on Earth." *Astrobiology* 19(4): 553–578. https://doi.org/10.1089/ast.2018.1926

Cavanagh, C. J. 2018. "Political Ecologies of Biopower: Diversity, Debates, and New Frontiers of Inquiry." *Journal of Political Ecology* 25: 350–425.

Chaikin, Andrew. 1998. *A Man on the Moon: The Voyages of the Apollo Astronauts.* Penguin.

Chakrabarty, Dipesh. 2007. *Provincializing Europe: Postcolonial Thought and Historical Difference*, new ed. Princeton University Press.

Chakrabarty, Dipesh. 2021. *The Climate of History in a Planetary Age.* University of Chicago Press.

Chang, David A. 2016. *The World and All the Things Upon It: Native Hawaiian Geographies of Exploration.* University of Minnesota Press.

Charbonneau, Rebecca. 2021. "Imaginative Cosmos: The Impact of Colonial Heritage in Radio Astronomy and the Search for Extraterrestrial Intelligence." *American Indian Culture and Research Journal* 45(1): 71–94.

Chinigò, Davide. 2019. "From the 'Merino Revolution' to the 'Astronomy Revolution': Land Alienation and Identity in Carnarvon, South Africa." *Journal of Southern African Studies* 45(4): 749–766.

Chojnacki, M., D. M. Burr, J. E. Moersch, and T. I. Michaels. 2011. "Orbital Observations of Contemporary Dune Activity in Endeavor Crater, Meridiani Planum, Mars." *Journal of Geophysical Research* 116: Article E00F19. https://doi.org/10.1029/2010JE003675

Clancey, William J. 2002. "Simulating 'Mars on Earth': A Report from FMARS Phase 2." In *On to Mars: Colonizing a New World Proceedings of the 1999, 2000 and 2001 Conventions of the Mars Society*, edited by Robert Zubrin and Frank Crossman. Apogee.

Clancey, William J. 2012. *Working on Mars: Voyages of Scientific Discovery with the Mars Exploration Rovers.* MIT Press.

Clark, Nigel. 2010. *Inhuman Nature: Sociable Life on a Dynamic Planet.* SAGE.

Clark, Nigel, and Bronislaw Szerszynski. 2021. *Planetary Social Thought: The Anthropocene Challenge to the Social Sciences.* Polity.

Clark, Timothy. 2012. "Scale." In *Telemorphosis: Theory in the Era of Climate Change*, edited by Tom Cohen, Vol. 1. Open Humanities Press.

Cockell, Charles S. 2007. *Space on Earth: Saving Our World by Seeking Others.* Macmillan.

Cockell, Charles S. 2014. "Types of Habitat in the Universe." *International Journal of Astrobiology* 13(2): 158–164. https://doi.org/10.1017/S1473550413000451

Cockell, Charles S. 2015. *Astrobiology: Understanding Life in the Universe.* Wiley-Blackwell.

Cockell, Charles S., Jesse P. Harrison, A. Stevens, et al. 2019. "A Low-Diversity Microbiota Inhabits Extreme Terrestrial Basaltic Terrains and Their Fumaroles: Implications for the Exploration of Mars." *Astrobiology* 19(3): 284–299.

Cockell, Charles S., Luke Hecht, and Hanna Landenmark. 2018. "Rapid Colonization of Artificial Endolithic Uninhabited Habitats." *International Journal of Astrobiology* 17(4): 386–401. https://doi.org/10.1017/S1473550417000398

Cohen, John M. 1969. *The Four Voyages of Columbus*. Century Hutchinson.

Colaprete, Anthony, Richard C. Elphic, Mark Shirley, et al. 2023. "Key Science Questions to Be Addressed by the Volatiles Investigating Polar Exploration Rover (VIPER)." In *54th Lunar and Planetary Science Conference, Held 13–17 March, 2023 at The Woodlands, Texas and Virtually*, LPI Contribution No. 2806. Lunar and Planetary Institute.

Collins, Samuel G. 2003. "Sail on! Sail on! Anthropology, Science Fiction, and the Enticing Future." *Science Fiction Studies* 30: 180–198.

Corfield, Richard. 2009. "Fossils on the Edge of Forever." *Astrobiology Magazine*, December. https://phys.org/news/2009-12-fossils-edge.html

Cosgrove, Denis. 2003. *Apollo's Eye: A Cartographic Genealogy of the Earth in the Western Imagination*. Johns Hopkins University Press.

Crary, Jonathan. 1992. *The Techniques of the Observer: On Vision and Modernity in the Nineteenth Century*. MIT Press.

Crichton, Michael. 1969. *The Andromeda Strain*. Knopf.

Danowski, Deborah, and Eduardo Viveiros de Castro. 2016. *The Ends of the World*. Polity.

Davis, Mike. 1998. *Ecology of Fear: Los Angeles and the Imagination of Disaster*. Holt.

Davis, Mike. 2005. *The Monster at Our Door: The Global Threat of Avian Flu*. New Press.

Dean, Jodi. 1998. *Aliens in America: Conspiracy Cultures from Outerspace to Cyberspace*. Cornell University Press.

Deans, Matthew, Jessica J. Marquez, Tamar Cohen, et al. 2017. "Minerva: User-Centered Science Operations Software Capability for Future Human Exploration." In *2017 IEEE Aerospace Conference*. IEEE.

Debaise, Didier. 2017. *Nature as Event. The Lure of the Possible*. Duke University Press.

de la Cadena, Marisol. 2015. *Earth Beings: Ecologies of Practice Across Andean Worlds*. Duke University Press.

Delaney, Carol. 2006. "Columbus's Ultimate Goal: Jerusalem." *Comparative Studies in Society and History* 48(2): 260–292.

DeLoughrey, Elizabeth M. 2013. "The Myth of Isolates: Ecosystem Ecologies in the Nuclear PACIFIC." *Cultural Geographies* 20(2): 167–184.

DeLoughrey, Elizabeth M. 2014. "Satellite Planetarity and the Ends of the Earth." *Public Culture* 26(2): 257–280.

De Musso, Federico. 2020. Voyage to Corona. *Medium*, June 18. https://medium.com/@federico.demusso/voyage-to-corona-a3d8a137115d

Denning, K., and A. Berea. 2019. "Figuring out, and Figuring in, the Human: Insights for Astrobiology from the Social Sciences." In *The Astrobiology Science Conference (AbSciCon) 2019, held in Seattle, WA, US, 24–28 March 2019, Session: Education, History and Community Development*. AGU.

Descola, Philippe, and Alessandro Pignocchi. 2022. *Ethnographies des Mondes à Venir.* Seuil.

De Smet, Elsa. 2018. *Voir l'Espace: Astronomie et Science Populaire Illustrée (1840–1969).* Presses Universitaires de Strasbourg.

Diamond, Jared. 1990. "New Zealand as an Archipelago: International Perspectives." In *Ecological Restoration of New Zealand Islands,* edited by D. R. Towns, Ch. Daugherty, and I. A. E. Atkinson. Conservation Sciences Publication.

Diamond, Judy, and Alan B. Bond. 1999. *Kea, Bird of Paradox. The Evolution and Behavior of a New Zealand Parrot.* University of California Press.

Dick, Steven J. 1991. "From the Physical World to the Biological Universe: Historical Developments Underlying SETI." In *Bioastronomy. Lecture Notes in Physics,* edited by J. Heidmann and M. J. Klein, Vol. 390. Springer.

Dick, Steven J. 2020. *Space, Time, and Aliens: Collected Works on Cosmos and Culture.* Springer.

Dick, Steven J., and James Edgar Strick. 2005. *The Living Universe: NASA and the Development of Astrobiology.* Rutgers University Press.

Dickens, Peter, and James Ormrod. 2007. *Cosmic Society: Towards a Sociology of the Universe.* Routledge.

Donnan, H., and T. M. Wilson. 1999. *Borders. Frontiers of Identity, Nation and State.* Berg.

Droser, Mary L., and James G. Gehling. 2015. "The Advent of Animals: The View from the Ediacaran." *Proceedings of the National Academy of Sciences of the USA* 112(16): 4865–4870.

Dubow, S. 2019. "200 Years of Astronomy in South Africa: From the Royal Observatory to the 'Big Bang' of the Square Kilometre Array." *Journal of Southern African Studies* 45(4): 663–687.

Dubrov, P. 2021. "Dark Green Capsicum Pods on the ISS." Instagram, photographed by Pyotr Dubrov (@Petr.v.dubrav) and Roscosmos official. November 3, 2021. https://www.instagram.com/p/CVz_X3fNVWY/?utm_source=ig_web_copy_link

Dunn, Alan. 1960. *Is There Intelligent Life on Earth? A Report to the Congress of Mars,* translated into English by the author. Simon & Schuster.

European Space Agency. 2019. "ESA's Mars Orbiters Did Not See Latest Curiosity Methane Burst." November 13, 2019. https://www.esa.int/Science_Exploration/Human_and_Robotic_Exploration/Exploration/ExoMars/ESA_s_Mars_orbiters_did_not_see_latest_Curiosity_methane_burst

European Space Agency. 2021. "Thomas Pasquet Cultivating Capsicum Flower on the ISS." Instagram, photographed by Thomas Pasquet (@Thom_Astro). October 9, 2021. https://www.instagram.com/p/CU0tW8MrwrB/?utm_source=ig_web_copy_link

Fairén, Alberto G., Alfonso F. Davila, Darlene Lim, et al. 2010. "Astrobiology Through the Ages of Mars: The Study of Terrestrial Analogues to Understand the Habitability of Mars." *Astrobiology* 10(8): 821–843. https://doi.org/10.1089/ast.2009.0440

Fais, G., A. Manca, F. Bolognesi, et al. 2022. "Wide Range Applications of Spirulina: From Earth to Space Missions." *Marine Drugs* 20(5): Article 299.

Faithorn, Lisa, and Baruch S. Blumberg. 2009. "Lessons Learned from the NASA Astrobiology Institute." In *Handbook of Research on Electronic Collaboration and Organizational Synergy*, edited by Janet Salmons and Lynn Wilson. IGI Global.

Finney, Ben, and Eric Jones, eds. 1985. *Interstellar Migration and the Human Experience.* University of California Press.

Formisano, Luciano. 1992. *Amerigo Vespucci. Letters from a New World.* Marsilio.

Fortey, Richard. 2005. *Earth: An Intimate History.* Vintage.

Freundlich, Martin, and Bernard E. Wagner, eds. 1967. *Exobiology: The Search for Extraterrestrial Life: Proceedings of an AAS/AAAS Symposium Held in New York, N.Y, December 30, 1967.* American Astronomical Society.

Garry, W. Brent, and Jacob E. Bleacher, eds. 2011. *Analogs for Planetary Exploration.* Geological Society of America.

Genovese, Taylor R. 2017. "The New Right Stuff: Social Imaginaries of Outer Space and the Capitalist Accumulation of the Cosmos." Master's thesis, Northern Arizona University.

Giovanni, Nikki. 2002. *Quilting the Black-Eyed Pea: Poems and Not Quite Poems.* Morrow.

Glaessner, Martin, and Mary Wade. 1971. "Praecambridium—A Primitive Arthropod." *Lethaia* 4(1): 71–77.

Good, Andrew, Alana Johnson, and Grey Hautaluoma. 2021. "NASA InSight's 'Mole' Ends Its Journey on Mars." NASA. January 14, 2021. http://www.nasa.gov/feature/jpl/nasa-insight-s-mole-ends-its-journey-on-mars

Goodman, Ronald. 1992. *Lakota Star Knowledge: Studies in Lakota Stellar Theology.* Sinte Gleska University Press.

Goodyear-ka'ōpua, Noelani. 2017. "Protectors of the Future, not Protestors of the Past: Indigenous Pacific Activism and Mauna a Wākea." *South Atlantic Quarterly* 116(1): 184–194.

Gordillo, Gastón. 2020. "Gravity: On the Primacy of Terrain." In *Voluminous States: Sovereignty, Materiality, and the Territorial Imagination*, edited by Franck Billé. Duke University Press.

Gorman, Alice C. 2005a. "The Archaeology of Orbital Space." Paper presented at the Australian Space Science Conference 2005.

Gorman, Alice C. 2005b. "The Cultural Landscape of Interplanetary Space." *Journal of Social Archaeology* 5(1): 85–107.

Gorman, Alice. 2017. "Pale Blue Dot: Everyday Material Culture on the International Space Station." Accessed August 28, 2020. https://doi.org/10.5284/1080986

Gorman, Alice. 2019. *Dr Space Junk vs The Universe: Archaeology and the Future.* MIT Press.

Gorman, Alice C., and Beth Laura O'Leary. 2013. "The Archaeology of Space Exploration." In *The Oxford Handbook of the Archaeology of the Contemporary World*, edited by Paul Graves-Brown, Rodney Harrison, and Angela Piccini. Oxford University Press.

Gough, Evan. 2019. "Slow Progress: NASA's Still Trying to Get InSight's Mole Working Again." July 5, 2019. https://phys.org/news/2019-07-nasa-insight-mole.html

Grafton, Anthony. 1992. *New Worlds, Ancient Texts: The Power of Tradition and the Shock of Discovery.* Belknap.

Grant, Edward. 1981. *Much Ado About Nothing: Theories of Space and Vacuum from the Middle Ages to the Scientific Revolution*. Cambridge University Press.

Gross, Matthias. 2007. "The Unknown in Process: Dynamic Connections of Ignorance, Non-Knowledge and Related Concepts." *Current Sociology* 55 (5): 742–759. https://doi.org/10.1177/0011392107079928

Gumbs, Alexis Pauline. 2018. *M Archive: After the End of the World*. Duke University Press.

Gunnarsson, Rafnar Orri, and Örlygur Hnefill Örlygsson, directors. 2019. *Cosmic Birth* [Film]. Colorwaves.

Guy, Jack. 2020. "NASA Fixes Mars Lander by Hitting It with a Shovel." CNN. March 19, 2020. https://www.cnn.com/2020/03/19/world/nasa-mars-lander-mission-scli-intl-scn/index.html

Hand, Kevin. 2020. *Alien Oceans: The Search for Life in the Depths of Space*. Princeton University Press.

Haraway, Donna J. 2013. *Simians, Cyborgs, and Women: The Reinvention of Nature*. Routledge.

Haraway, Donna J. 2016. *Staying with the Trouble: Making Kin in the Chthulucene*. Duke University Press.

Harjo, Laura. 2019. *Spiral to the Stars: Mvskoke Tools of Futurity*. University of Arizona Press.

Harman, Graham. 2009. *Prince of Networks: Bruno Latour and Metaphysics*. re.press.

Harman, Graham. 2018. *Object-Oriented Ontology: A New Theory of Everything*. Penguin.

Harman, Graham. 2019. *Immaterialism: Objects and Social Theory*. Wiley. https://www.perlego.com/book/1536331/immaterialism-objects-and-social-theory-pdf

Harvey, David. 2000. *Spaces of Hope*. University of California Press.

Hau'Ofa, Epeli. 1994. "Our Sea of Islands." *The Contemporary Pacific* 6(1): 2–17.

Heidegger, Martin. 1993. *Basic Writings*. Routledge.

Heidegger, Martin. 2008a. *Towards the Definition of Philosophy*. Athlone Press.

Heidegger, Martin. 2008b. *Being and Time*. HarperCollins.

Helmreich, Stefan. 2006. "The Signature of Life: Designing the Astrobiological Imagination." *Grey Room* (23): 66–95.

Helmreich, Stefan. 2009. *Alien Ocean: Anthropological Voyages in Microbial Seas*. University of California Press.

Helmreich, Stefan. 2012. "Extraterrestrial Relativism." *Anthropological Quarterly* 85(4): 1125–1139.

Helmreich, Stefan 2016. *Sounding the Limits of Life: Essays in the Anthropology of Biology and Beyond*. Princeton University Press.

Hersch, Matthew H. 2012. *Inventing the American Astronaut*. Palgrave Macmillan.

Hesse, Mary B. 1966. *Models and Analogy in Science*. Revised ed. Notre Dame University Press.

Hill, Jordan R., Barrett S. Caldwell, Michael Downs, Michael J. Miller, and Darlene S. S. Lim. 2019. "Remote Physiological Monitoring in a Mars Analog Field Setting." *IISE Transactions on Healthcare Systems Engineering* 8(3): 227–236.

Hobart, Hi'ilei Julia. 2019. "At Home on the Mauna: Ecological Violence and Fantasies of Terra Nullius on Maunakea's Summit." *Native American and Indigenous Studies* 6(2): 30–50.

Hoeppe, Götz. 2012. "Astronomers at the Observatory: Place, Visual Practice, Traces." *Anthropological Quarterly* 85(4): 1141–1160.

Hoffman, G. Lyle, N. Y. Lu, E. E. Salpeter, et al. 1993. "Arecibo Mapping of Three Extended H I Galaxy Disks and Conjectures on What Lies Beyond." *Astronomical Journal* 106: 39–55.

Hofstadter, Douglas R., and Emmanuel Sander. 2013. *Surfaces and Essences: Analogy as the Fuel and Fire of Thinking.* Basic.

Holbrook, Jarita, Diane Gu, and Sharon Traweek. 2018. "Reaching for the Stars Without an Invitation." *American Scientist* 106(5): 296–298.

Holyoak, Keith J., and Paul Thagard. 1995. *Mental Leaps: Analogy in Creative Thought.* MIT Press.

Hoshida, George Yoshio. 1981. "George Yoshio Hoshida Oral History Interview." JCCH-51327-OH. Japanese Cultural Center of Hawai'i.

Houdart, Sophie, and Christine Jungen. 2015. "Cosmos Connections." *Gradhiva, Revue d'anthropologic et d'histoire des arts* 22: 4–23.

Howell, Elizabeth. 2020. "Mars Lander Reveals New Details About the Red Planet's Strange Magnetic Field." Space.Com. February 26. https://www.space.com/mars-weird-magnetic-field-insight-lander-discovery.html

Hughes, Scott S., Christopher W. Haberle, Shannon Kobs Nawotniak, et al. 2019. "Basaltic Terrains in Idaho and Hawai'i as Planetary Analogs for Mars Geology and Astrobiology." *Astrobiology* 19(3): 260–283.

Ingold, Tim. 2011. *Being Alive: Essays on Movement, Knowledge and Description.* Routledge.

Ingold, Tim, and Terhi Kurttila. 2000. "Perceiving the Environment in Finnish Lapland." *Body & Society* 6(3–4): 183–196. https://doi.org/10.1177/1357034X00006003010

Jacka, Jerry K. 2018. "The Anthropology of Mining: The Social and Environmental Impacts of Resource Extraction in the Mineral Age." *Annual Review of Anthropology* 47(1): 61–77.

Jameson, Fredric. 2007. *Archaeologies of the Future: The Desire Called Utopia and Other Science Fictions.* Verso.

Jasanoff, Sheila, and Marybeth Long Martello. 2004. "Heaven and Earth: The Politics of Environmental Images." In *Earthly Politics: Local and Global in Environmental Governance*, edited by Sheila Jasanoff and Marybeth Long Martello. MIT Press.

Jeevendrampillai, David, and Aaron Parkhurst. 2021. "Making a Martian Home: Finding Humans on Mars Through Utopian Architecture." *Home Cultures* 18(1): 25–46.

Jensen, S., and O. Zenker. 2015. "Homelands as Frontiers: Apartheid's Loose Ends—An Introduction." *Journal of Southern African Studies* 41(5): 937–952.

Jobson, Ryan Cecil. 2020. "The Case for Letting Anthropology Burn: Sociocultural Anthropology in 2019." *American Anthropologist* 122(2): 259–271.

Johnson, Kirsten. 2016. "BASALT Project Paves Way for Future Mission on Mars." *Hawaii Tribune-Herald*, November 15, 2016.

Jones, Graham M. 2018. *Magic's Reason: An Anthropology of Analogy*. University of Chicago Press.

Kafer, Alison. 2013. *Feminist, Queer, Crip*. Indiana University Press.

Kaplan, Sarah. 2016. "How NASA Is Rehearsing for a Mission to Mars." *The Washington Post*, December 20.

Kattenhorn, Simon A. and Louise M. Prockter. 2014. "Evidence for Subduction in the Ice Shell of Europa." *Nature Geoscience* 7: 762–767.

Kaufman, Marc. 2022 "Life Here and Beyond." NASA. Last modified October 12, 2022. https://astrobiology.nasa.gov/about

Keane, Webb. 2005. "Signs Are not the Garb of Meaning: On the Social Analysis of Material Things." In *Materiality*, edited by Danny Miller. Duke University Press.

Keane, Webb. 2007. *Christian Moderns: Freedom and Fetish*. Duke University Press.

Keeton, Kathryn E., Alexandra Whitemire, Alan H. Feiveson, Robert Ploutz-Snyder, Lauren B. Leveton, and Camille Shea. 2011. "Analog Assessment Tool Report." NASA/TP–2011-216146. NASA.

Kiang, Nancy Y. 2008. "The Color of Plants on Other Worlds." *Scientific American* 298(4): 48–55.

Kīlauea. 2018. "2018 Lower East Rift Zone Eruption and Summit Collapse." Blog. U.S. Geological Survey.

Kinchy, Abby, Roopali Phadke, and Jessica M. Smith. 2018. "Engaging the Underground: An STS Field in Formation." *Engaging Science, Technology, and Society* 4: 22–42.

Kite, Suzanne. 2021. "'What's on the Earth Is in the Stars; and What's in the Stars Is on the Earth': Lakota Relationships with the Stars and American Relationships with the Apocalypse." *American Indian Culture and Research Journal* 45(1): 137–156.

Klein, JoAnna. 2018. "The Genes That Make Parrots into the Humans of the Bird World." *The New York Times*, December 7.

Klinger, Julie. M. 2018. *Rare Earth Frontiers: From Terrestrial Subsoils to Lunar Landscapes*. Cornell University Press.

Klinger, Julie M. 2019. "Environmental Geopolitics and Outer Space." *Critical Geopolitics of Outer Space* 26(3): 666–703.

Klinger, Julie M. 2021. "Environmental Geopolitics and Outer Space." *Geopolitics* 26(3): 666–703. doi:10.1080/14650045.2019.1590340

Knoll, Harry. 2004. *Life on a Young Planet: The First Three Billion Years of Evolution on Earth*. Princeton University Press.

Kohler, Robert E. 2002. *Landscapes and Labscapes Exploring the Lab–Field Border in Biology*. University of Chicago Press.

Kohn, Eduardo. 2013. *How Forests Think: Toward an Anthropology Beyond the Human*. University of California Press.

Kopytoff, Igor 1987. "The Internal African Frontier—The Making of African Political Culture." In *The African Frontier—The Reproduction of Traditional African Societies*, edited by I. Kopytoff. Indiana University Press.

Korf, B., and T. Raeymaekers. 2013. "Introduction: Border, Frontier and the Geography of Rule at the Margins of the State." In *Violence on the Margins*, edited by B. Korf and T. Raeymaekers. Palgrave Macmillan.

Korf, B., T. Hagmann, and R. Emmenegger. 2015. "Re-Spacing African Drylands: Territorialization, Sedentarization and Indigenous Commodification in the Ethiopian Pastoral Frontier." *Journal of Peasant Studies* 42(5): 881–901.

Kusch, Martin. 2002. *Knowledge by Agreement: The Programme of Communitarian Epistemology*. Oxford University Press.

Laflamme, Marc, Shuhai Xiao, and Michal Kowalewski. 2009. "Osmotrophy in Modular Ediacara Organisms." *Proceedings of the National Academy of Sciences of the USA* 106(34): 14438–14443.

Lakoff, George, and Mark Johnsen. 2013. *Metaphors We Live By*. University of Chicago Press.

Lanza, Nina L., R. C. Wiens, S. M. Clegg, et al. 2010. "Calibrating the ChemCam Laser-Induced Breakdown Spectroscopy Instrument for Carbonate Minerals on Mars." *Applied Optics* 49(13): C211–C117.

Larson, William E., Martin Picard, Gerald B. Sanders, et al. 2012. "NASA's Lunar Polar Ice Prospector, RESOLVE." NASA. https://ntrs.nasa.gov/api/citations/20120017905/downloads/20120017905.pdf

Latour, Bruno. 1996. *Aramis, or The Love of Technology*, translated by Catherine Porter. Harvard University Press.

Latour, Bruno. 2012. *We Have Never Been Modern*. Harvard University Press.

Latour, Bruno. 2018. *Down to Earth: Politics in the New Climatic Regime*. Polity.

Latour, Bruno. 2019. *Down to Earth: Politics in the New Climatic Regime*. Polity.

Latour, Bruno, and Peter Weibel. 2020. *Critical Zones: The Science and Politics of Landing on Earth*. MIT Press.

Lazier, Benjamin. 2011. "Earthrise; or, The Globalization of the World Picture." *American Historical Review* 116(3): 602–630.

Lee, Pascal. 2002. "Mars on Earth: The NASA Haughton-Mars Project." *Ad Astra*, June.

Legassick, M. 1972. "The Frontier Tradition in South African Historiography." In *Collected Seminar Papers*. Institute of Commonwealth Studies.

Lempert, William. 2021. "From Interstellar Imperialism to Celestial Wayfinding: Prime Directives and Colonial Time-Knots in SETI." *American Indian Culture and Research Journal* 45(1): 45–70.

Lepselter, Susan. 2016. *The Resonance of Unseen Things: Poetics, Power, Captivity, and UFOs in the American Uncanny*. University of Michigan Press.

Levinas, Emmanuel. 1990. "Heidegger, Gagarin and Us." In *Difficult Freedom: Essays on Judaism*, translated by Seán Hand. Johns Hopkins University Press.

Lewin, Roger. 1984. "Alien Beings Here on Earth." *Science* 223(4631): 39.

Lewis, Cathleen. 2012. "Okudzhava and Scott-Heron: The Social Critique Soundtrack of the Space Race." Paper presented at Sounds of Space workshop, Freie Universität Berlin, Berlin, Germany, November 30–December 1.

Lim, Darlene S. S., Andrew F. J. Abercromby, Shannon E. Kobs Nawotniak, et al. 2019. "The BASALT Research Program: Designing and Developing Mission Elements in Support of Human Scientific Exploration of Mars." *Astrobiology* 19(3): 245–259.

Lim, Darlene S. S., Allyson L. Brady, Andrew F. Abercromby, et al. 2011. "A Historical Overview of the Pavilion Lake Research Project—Analog Science and Exploration in an Underwater Environment." In *Analogs for Planetary Exploration*, edited by W. Brent Garry and Jacob E. Bleacher, Vol. 483. Geological Society of America.

Lim, Darlene S. S., Zara Mirmalek, Anthony Colaprete, et.al. 2024. "VIPER Science Operations: Science Traverse Planning, Perspectives, Processes and Tools." In *55th Lunar and Planetary Science Conference, Held 11–15 March, 2024 at The Woodlands, Texas/Virtual.* Lunar and Planetary Institute.

Lim, Darlene S. S., Zara Mirmalek, Nicole Raineault, David S. Lees, and Chris German. 2021. "SUBSEA—A Joint NASA–NOAA Planetary and Ocean Sciences Analog Research Program." In *Workshop on Terrestrial Analogs for Planetary Exploration*, No. 2595. Lunar and Planetary Institute.

Limbert, Robert W. 1924. "Among the 'Craters of the Moon.'" *National Geographic Magazine.*

Lopes, Rosaly, and Michael Carroll. 2008. *Alien Volcanoes.* Johns Hopkins University Press.

López, Alejandro Martín. 2011. "Ethnoastronomy as an Academic Field: A Framework for a South American Program." *Proceedings of the International Astronomical Union* 7(S278): 38–49.

Lorenz, Ralph, and James Zimbelman. 2014. *Dune Worlds: How Windblown Sand Shapes Planetary Landscapes.* Springer.

Luedemann, Gary. 1989. "Procedure for X-Ray Fluorescence Analysis." TWS-ESS-DP-111. U.S. Nuclear Regulatory Institution.

MacDonald, Fraser. 2007. "Anti-Astropolitik: Outer Space and the Orbit of Geography." *Progress in Human Geography* 31(5): 592–615.

Made In Space. 2020. "Space Station's Commercial 3D Printer Makes Its 1st Tool." Accessed October 29, 2020. https://www.space.com/33166-space-station-commercial-3d-printer-first-tool-photos.html

Maile, David Uahikeaikalei 'ohu. 2021. "On Being Late: Cruising Mauna Kea and Unsettling Technoscientific Conquest in Hawai'i." *American Indian Culture and Research Journal* 45(1): 95–122.

Marcheselli, Valentina. 2019. "Life as-We-Don't-Know-It: Research Repertoires and the Emergence of Astrobiology." Doctoral dissertation, University of Edinburgh.

Marcheselli, Valentina. 2020. "The Shadow Biosphere Hypothesis: Non-Knowledge in Emerging Disciplines." *Science, Technology, & Human Values* 45(4): 636–658. https://doi.org/10.1177/0162243919881207

Marcheselli, Valentina. 2022. "The Exploration of the Earth Subsurface as a Martian Analogue." *Tecnoscienza: Italian Journal of Science & Technology Studies* 13(1): 22–45. https://doi.org/10.6092/ISSN.2038-3460/17563

Mark, Gloria, Steve Abrams, and Nayla Nassif. 2003. "Group-to-Group Distance Collaboration: Examining the 'Space Between.'" In *The Proceedings of the 8th European Conference of Computer-Supported Cooperative Work (ECSCW'03)*. Kluwer.

Marlow, Jeffrey J., Zita Martins, and Mark A. Sephton. 2008. "Mars on Earth: Soil Analogues for Future Mars Missions." *Astronomy and Geophysics* 49(2): 2.20–2.23. https://doi.org/10.1111/j.1468-4004.2008.49220.x

Martins, Zita, Hervé Cottin, Julia Michelle Kotler, et al. 2017. "Earth as a Tool for Astrobiology—A European Perspective." *Space Science Reviews* 209(1–4): 43–81. https://doi.org/10.1007/s11214-017-0369-1

Maruyama, Magorah, and Arthur M. Harkins, eds. 1975. *Cultures Beyond Earth*. Vintage Books.

Mavhunga, C. C. 2017. *What Do Science, Technology, and Innovation Mean from Africa?* MIT Press.

Mavimbela, V. 1998. "The African Renaissance: A Workable Dream." In *South Africa and Africa: Reflections on the African Renaissance*, edited by G. Pere, A. Nieuwkerk, and K. Lambrechts, FGD Occasional Paper No. 17. Foundation for Global Dialogue.

Mbembe, A. 2019. "Strategic Narratives of Technology and Africa." Conference keynote, presented at the Strategic Narratives of Technology and Africa Conference, Funchal, September 2, 2017, Madeira Interactive Technologies Institute.

McGoey, Linsey. 2012. "Strategic Unknowns: Towards a Sociology of Ignorance." *Economy and Society* 41(1): 1–16. https://doi.org/10.1080/03085147.2011.637330

McKay, Christopher P., E. Imre Friedmann, Benito Gómez-Silva, Luis Cáceres-Villanueva, Dale T. Andersen, and Ragnhild Landheim. 2004. "Temperature and Moisture Conditions for Life in the Extreme Arid Region of the Atacama Desert: Four Years of Observations Including the El Niño of 1997–1998." *Astrobiology* 3(2): 393–406. https://doi.org/10.1089/153110703769016460

McMenamin, Mark. 1986. "The Garden of Ediacara." *Palaios* 1(2): 178–182.

McMenamin, Mark A. S., and Dianna L. S. McMenamin. 1993. *Hypersea: Life on Land*. Columbia University Press.

Meadows, Donella H, Dennis L. Meadows, Jørgen Randers, and William W. Behrens, III. 1972. *The Limits to Growth; A Report for the Club of Rome's Project on the Predicament of Mankind*. Universe Books.

Meir, J. Mizuna 2019. "Lettuce Growing in V.E.G.G.I.E." Instagram, photographed by Jessica Meir (@astro_jessica). October 24, 2019.

Mendenhall, Elizabeth. 2018. "Treating Outer Space like a Place: A Case for Rejecting Other Domain Analogies." *Astropolitics* 16(2): 97–118. https://doi.org/10.1080/14777622.2018.1484650

Merron, J., and S. Lamoureaux. 2024. "Electromagnetic Economies of Worth: Repurposing a Radio Dish and Debating Technoscientific Modernity at the Equator." *Environment and Planning D: Society and Space*, forthcoming.

Messeri, Lisa. 2014. "Earth as Analog: The Disciplinary Debate and Astronaut Training That Took Geology to the Moon." *Astropolitics* 12(2–3): 196–209.

Messeri, Lisa. 2016. *Placing Outer Space: An Earthly Ethnography of Other Worlds*. Duke University Press.

Messeri, Lisa. 2017. "Resonant Worlds: Cultivating Proximal Encounters in Planetary Science." *American Ethnologist* 44(1): 131–142.

Miller, Michael J., Matthew J. Miller, Delia Santiago-Materese, Marc A. Seibert, and Darlene S. S. Lim. 2019. "A Flexible Telecommunication Architecture for Human Planetary Exploration Based on the BASALT Science-Driven Mars Analog." *Astrobiology* 19(3): 478–496.

Miller, Patricia C. 1994. "Desert Ascetisicm and 'the Body from Nowhere.'" *Journal of Early Christian Studies* 2(2): 137–153.

Miller, Patricia C. 2009. *The Corporeal Imagination: Signifying the Holy in Late Ancient Christianity*. University of Pennsylvania Press.

Mirmalek, Zara. 2017. "Inspiring Innovation: On Low-Tech in High-Tech Development." *Interactions* 24(4): 50–55.

Mirmalek, Zara. 2020. *Making Time on Mars*. MIT Press.

Mirmalek, Zara, Darlene S. S. Lim, and Anthony Colaprete. 2021. "Remote Science Work Support, Context, and Approach on NASA's VIPER Mission." In *52nd Lunar and Planetary Science Conference 2021*, No. 1734. Lunar and Planetary Institute.

Mirmalek, Zara, Darlene S. S. Lim, and Anthony Colaprete. 2024. "Work Ethnography Applied for Science Operations on NASA's VIPER Mission." Society for Applied Anthropology.

Mirmalek, Zara, Matthew J. Miller, and Darlene S. S. Lim. 2019. "Envisioning an Analog Work Domain Across Deep-Ocean Exploration and Human–Robot Spaceflight Exploration." In *Proceedings of the Human Factors and Ergonomics Society 2019 Annual Meeting*. Human Factors and Ergonomics Society.

Mitchell, Sean T. 2017. *Constellations of Inequality: Space, Race, and Utopia in Brazil*. University of Chicago Press.

Moffitt Watts, Pauline. (1985). "Prophecy and Discovery: On the Spiritual Origins of Christopher Columbus's 'Enterprise of the Indies." *American Historical Review* 90(1): 73–102.

Mondzain, Marie-José. 2004. *Image, Icon, Economy: The Byzantine Origins of the Contemporary Imaginary*. Stanford University Press.

Morgagni, Simone. 2012. "Affordances as Possible Actions: Elements for a Semiotic Approach." In *Proceedings of the 10th World Congress of the International Association for Semiotic Studies*. International Association for Semiotic Studies.

Morton, Timothy. 2013. *Hyperobjects: Philosophy and Ecology After the End of the World*. University of Minnesota Press.

Munn, Nancy. 1977. "The Spatiotemporal Transformations of Gawa Canoes." *Journal de la Société des Océanistes* 33(54–55): 39–53.

Munn, Nancy. 1986. *The Fame of Gawa: A Symbolic Study of Value Transformation in a Massim Society*. Cambridge University Press.

Myers, Natasha. 2019. "From Edenic Apocalypse to Gardens Against Eden: Plants and People in and After the Anthropocene." In *Infrastructure, Environment and Life in the Anthropocene*, edited by Kregg Hetherington. Duke University Press.

Nardi, Bonnie A. 2005. "Beyond Bandwidth: Dimensions of Connection in Interpersonal Communication." *Computer Supported Cooperative Work* 14(2): 91–130.

Nichols, Robert. 2013. "Indigeneity and the Settler Contract Today." *Philosophy & Social Criticism* 39(2): 165–186.

Nolen, Stephanie. 2002. *Promised the Moon*. Thunder's Mouth Press.

Nscrypt. 2020. "Nscrypt Has 3D Bioprinted a Human Knee Meniscus in Space." Accessed October 29, 2020. https://3dprintingindustry.com/news/nscrypt-has-3d-bioprinted-a-human-knee-meniscus-in-space-171627

Nygren, A., and S. Rikoon. 2008. "Political Ecology Revisited: Integration of Politics and Ecology Does Matter." *Society and Natural Resources* 21(9): 767–782.

Oliver, Kelly. 2015. *Earth and World: Philosophy After the Apollo Missions.* Columbia University Press.

Olson, Valerie. 2012. "Political Ecology in the Extreme: Asteroid Activism and the Making of an Environmental Solar System." *Anthropological Quarterly* 85(4): 1027–1044.

Olson, Valerie. 2013. "NEOspace: The Solar System's Emerging Environmental History and Politics." In *New Natures: Joining Environmental History with STS*, edited by Dolly Jörgensen, Finn-Arne Jörgensen, and Sara Pritchard. Pittsburgh University Press.

Olson, Valerie. 2018. *Into the Extreme. U.S. Environmental Systems and Politics Beyond Earth.* University of Minnesota Press.

Olson, Valerie. 2023. "On Paradoxical Ground: Refielding Social Studies." In *The Routledge Handbook of Outer Space Studies*, edited by J. F. Salazar and A. Gorman. Routledge.

Olson, Valerie, & Lisa Messeri. 2015. "Beyond the Anthropocene: Un-Earthing an Epoch." *Environment and Society* 6(1): 28–47.

Olympus. 2022. "Portable GPS-XRF Protocols for Field Archaeology." *Olympus Industrial Resources Application Notes.* Blog.

Omarov, Tuken B., and Bulat T. Tashenov. 2005. "Tikhov's Astrobotany as a Prelude to Modern Astrobiology." In *Perspectives in Astrobiology*, edited by Richard B. Hoover, Alexei Yu. Rozanov, and Roland Paepe, Vol. 366. NATO.

O'Neill, Gerard K. 1976. *The high frontier: Human colonies in space.* Morrow.

O'Reilly, Jessica. 2016. "Sensing the Ice: Field Science, Models, and Expert Intimacy with Knowledge." *Journal of the Royal Anthropological Institute* 22(1): 27–45.

Ormrod, James S. 2009. "Phantasy and Social Movements: An Ontology of Pro-Space Activism." *Social Movement Studies* 8(2): 115–129.

Ormrod, James S., and Peter Dickens. 2016. *The Palgrave Handbook of Society, Culture and Outer Space.* Palgrave Macmillan.

Paine, Robert. 1995. "Columbus and Anthropology and the Unknown." *Journal of the Royal Anthropological Institute* 1(1): 47–65.

Painter, Fantasia. 2021. "G-Men, Green Men, and Red Land: Extraterrestrial Miscreants, Federal Jurisdiction, and Exceptional Space." *American Indian Culture and Research Journal* 45(1), 123–136.

Pálsson, Gísli. 2007. *Anthropology and the New Genetics.* Cambridge University Press.

Parkington, J., D. Morris, and J. M. de Prada-Samper. 2019. "Elusive Identities: Karoo|Xam Descendants and the Square Kilometre Array." *Journal of Southern African Studies* 45(4): 729–747.

Payler, Samuel J., Zara Mirmalek, Scott Hughes, et al. 2019. "Developing Intra-EVA Science Support Team Practices for a Human Mission to Mars." *Astrobiology* 19(3): 387–400.

Peluso, N., and C. Lund. 2011. "New Frontiers of Land Control: Introduction." *Journal of Peasant Studies* 38(4): 667–681.

Penley, Constance. 1997. *Nasa/Trek: Popular Science and Sex in America.* Verso.

Penn, Nigel. 2005. *The Forgotten Frontier: Colonist and Khoisan on the Cape's Northern Frontier in the 18th Century*. Ohio University Press.

Peterson, Kristin, and Valerie Olson. 2024. *The Ethnographer's Way: A Handbook for Multidimensional Research Design*. Duke University Press.

Pettit, Don. 2012. "Diary of a Space Zucchini. Letters to Earth." Blog. NASA. April 3. https://blogs.nasa.gov/letters/2012/04/03/post_1333471169633

Pietz, William. 1985. "The Problem of the Fetish, I." *Res* 9: 5–17.

Pietz, William. 1987. "The Problem of the Fetish, II." *Res* 13: 23–45.

Pitrou, Perig. 2019. "Figuration, Configuration, Reconfiguration: Le Vivant Entre Métonymie et Métaphore." *Cahiers d'Anthropologie Sociale* 19: 13–33.

Pitrou, Perig. 2020. "Le Biomimétisme Comme Système." *Techniques & Culture* 73: 34–43.

Popper, J., and W. Berry. 2021. "Fragmenting a Monolith: Exploring and Disrupting an Outer Space Imaginary." Doctoral dissertation, Universität für Künsterische und Industrielle Gestaltung Linz, Austria.

Poulet, F., J.-P. Bibring, J. F. Mustard, et al. 2005. "Phyllosilicates on Mars and Implications for Early Martian Climate." *Nature* 438(7068): 623–627.

Povinelli, Elizabeth A. 2016. *Geontologies: A Requiem to Late Liberalism*. Duke University Press.

Praet, Istvan. 2013. *Animism and the Question of Life*. Routledge.

Praet, Istvan. 2017. "Astrobiology and the Ultraviolet World." *Environmental Humanities* 9(2): 378–397.

Praet, Istvan, and Juan Francisco Salazar. 2017. "Introduction: Familiarizing the Extraterrestrial/Making Our Planet Alien." *Environmental Humanities* 9(2): 309–324.

Prescod-Weinstein, Chanda. 2020. "Making Black Women Scientists Under White Empiricism: The Racialization of Epistemology in Physics." *Signs: Journal of Women in Culture and Society* 45(2): 421–447.

Prescod-Weinstein, Chanda. 2021. *The Disordered Cosmos: A Journey into Dark Matter, Spacetime, and Dreams Deferred*. Bold Type Books.

Preston, Luisa J., Simon Barber, and Monica Grady. 2023. "CAFE—Concepts for Activities in the Field for Exploration—TN2: The Catalogue of Planetary Analogues." European Space Agency. Accessed April 4, 2024. http://esamultimedia.esa.int/docs/gsp/The_Catalogue_of_Planetary_Analogues.pdf

Pyne, Stephen. 2007. "The Extraterrestrial Earth: Antarctica as Analogue for Space Exploration." *Space Policy* 23(3): 147–149.

Radin, Joanna. 2019. "The Speculative Present: How Michael Crichton Colonized the Future of Science and Technology." *Osiris* 34(1): 297–315.

Rand, Lisa Ruth. 2019. "Falling Cosmos: Nuclear Reentry and the Environmental History of Earth Orbit." *Environmental History* 24(1): 78–103.

Redfield, Peter. 2000. *Space in the Tropics: From Convicts to Rockets in French Guiana*. University of California Press.

Redfield, Peter. 2002. "The Half-Life of Empire in Outer Space." *Social Studies of Science* 32(5–6): 791–825.

Reid, Ann, and Merry Buckley. 2011. *The Rare Biosphere*. American Academy of Microbiology. https://www.ncbi.nlm.nih.gov/books/NBK560451/pdf/Bookshelf_NBK560451.pdf

Reid, Lauren. 2023. "Divergent Extraterrestrial Worlds: Navigating Cosmo-Practices on Two Mountaintops in Thailand." In *The Routledge Handbook of Social Studies of Outer Space*, edited by Juan Francisco Salazar and Alice Gorman. Routledge.

Republic of South Africa. 1996. "White Paper on Science & Technology: Preparing for the 21st Century." Department of Arts, Culture, Science and Technology, September 4.

Republic of South Africa. 2007. "Astronomy Geographic Advantage Act. Act 21 of 2007." Government Gazette.

Republic of South Africa. 2020. "Declaration of Certain Properties Situated in the Northern Cape Province the Meerkat National Park." Proclamation 15 of 2020, No. 43145.

Rifkin, Mark. 2017. *Beyond Settler Time: Temporal Sovereignty and Indigenous Self-Determination*. Duke University Press.

Riskin, Jessica. 2016. *The Restless Clock: A History of the Centuries-Long Argument over What Makes Living Things Tick*. Chicago University Press.

Robinson, Kim Stanley. 1993. *Red Mars*. Spectra.

Robinson, Kim Stanley. 1997. *Blue Mars*. Spectra.

Ruhleder, Karen, and Brigitte Jordan. 2001. "Co-Constructing Non-Mutual Realities: Delay-Generated Trouble in Distributed Interaction." *Computer Supported Cooperative Work* 10(1): 113–138.

Russell, Michael J., Alison E. Murray, and Kevin P. Hand. 2017. "The Possible Emergence of Life and Differentiation of a Shallow Biosphere on Irradiated Icy Worlds: The Example of Europa." *Astrobiology* 17(12): 1265–1273. https://doi.org/10.1089/ast.2016.1600

Salazar, Juan Francisco. 2017a. "Microbial Geographies at the Extremes of Life." *Environmental Humanities* 9(2): 398–417.

Salazar, Juan Francisco. 2017b. "Antarctica and Outer Space: Relational Trajectories." *The Polar Journal* 7(2): 259–269.

Salazar, Juan Francisco. 2018. "Ice Cores as Temporal Probes." *Journal of Contemporary Archaeology* 1(5): 32–43.

Salazar, Juan Francisco, Céline Granjou, Matthew Kearnes, Anna Krzywoszynska, and Manuel Tironi, eds. 2020. *Thinking with Soils: Material Politics and Social Theory*. Bloomsbury.

Sammler, Katherine G., and Casey R. Lynch. 2021a. "Spaceport America: Contested Offworld Access and the Everyman Astronaut." *Geopolitics 26*(3): 704–728.

Sammler, Katherine G., and Casey R. Lynch. 2021b. "Apparatuses of Observation and Occupation: Settler Colonialism and Space Science in Hawai'i." *Environment and Planning D: Society and Space* 39(5): 945–965.

Schaffer, Simon. 2007. "Astrophysics, Anthropology and Other Imperial Pursuits." In *Anthropology and Science: Epistemologies in Practice*, edited by Jeanette Edwards, Penny Harvey, and Peter Wade. Routledge.

Scharmen, Fred. 2019a. "Jeff Bezos Dreams of a 1970s Future." Bloomberg UK, May 13. https://www.bloomberg.com/news/articles/2019-05-13/why-jeff-bezos-s-space-habitats-already-feel-stale

Scharmen, Fred. 2019b. *Space Settlements.* Columbia University Press.

Schultz, Nikolaj. 2020. "Life as Exodus." In *Critical Zones: The Science and Politics of Landing on Earth*, edited by Bruno Latour and Peter Weibel. MIT Press.

Scoles, Sarah. 2020. "The Doctor from Nazi Germany and the Search for Life on Mars." *The New York Times*, July 24: https://www.nytimes.com/2020/07/24/science/mars-jars-strughold.html

Scott-Heron, Gil. 1970. "Whitey on the Moon." On *Small Talk at 125th and Lenox.* Musical recording. Flying Dutchman Records.

Sehlke, Alexander, Zara Mirmalek, David Burtt, et al. 2019. "Requirements for Portable Instrument Suites During Human Scientific Exploration of Mars." *Astrobiology* 19(3): 401–425.

Sehlke, Alexander, Zara Mirmalek, Barbara Cohen, et al. 2017. "The Ultimate Geologic Tricorder? Handheld Science Instruments and Requirements for Future Human Exploration Missions on Other Worlds." Paper presented at the 48th Lunar and Planetary Science Conference, The Woodlands, Texas, March 20.

Sehlke, Alexander, Zara Mirmalek, Jennifer Heldmann, and Darlene S. S. Lim. 2022. "Requirements of Handheld VNIR and XRF During Extra Vehicular Activities." Paper presented at the Lunar Petrology and Landed Instruments Interchange Workshop, Jet Propulsion Laboratory, Pasadena, CA, August 9.

Seilacher, Adolf. 1989. "Vendozoa: Organismic Construction in the Proterozoic Biosphere." *Lethaia* 22: 229–239.

Seilacher, Adolf, Dmitri Grazhdankin, and Anton Legouta. 2003. "Ediacaran Biota: The Dawn of Animal Life in the Shadow of Giant Protists." *Paleontological Research* 7(1): 43–54.

Selvans, Michelle M. 2014. "Plate Tectonics on Ice." *Nature Geoscience* 7: 695–696.

Shammas, Victor L., and Tomas B. Holen. 2019. "One Giant Leap for Capitalistkind: Private Enterprise in Outer Space." *Palgrave Communications* 5(1): 10.

Sharpe, Colin Q. 2016. "Slavery, Migration and Local Development in the Western US." Master's dissertation, City University of New York.

Shiga, S. 2008. "Stephen Hawking Calls for Moon and Mars Colonies." *NewScientist*, April 21. https://www.newscientist.com/article/dn13748-stephen-hawking-calls-for-moon-and-mars-colonies

Shipley, J. W. 2010. "Africa in Theory: A Conversation Between Jean Comaroff and Achille Mbembe." *Anthropological Quarterly* 83(3): 653–678.

Shorter, David, and Kim TallBear. 2021. "An Introduction to Settler Science and the Ethics of Contact." *American Indian Culture and Research Journal* 45(1), 1–8.

Siddiqi, Asif A. 2000. *Challenge to Apollo: The Soviet Union and the Space Race, 1945–1974.* Vol. 4408. NASA History Division, Office of Policy and Plans.

Siddiqi, Asif A. 2010. *The Red Rockets' Glare: Spaceflight and the Russian Imagination, 1857–1957.* Cambridge University Press.

Siebert, Lee. 2019. "When Volcanoes Fall Down: Catastrophic Collapse and Debris Avalanches." U.S. Geological Survey. https://pubs.usgs.gov/fs/2019/3023/fs20193023_v1.2.pdf

Sierhaus, Maarten, M. H. Sims, William J. Clancey, and P. Lee. 2000. "Applying Multiagent Simulation to Planetary Surface Operations." Paper presented at the COOP'2000 Workshop on Modelling Human Activity, Sophia Antipolis, France.

Simpson, Audra. 2014. *Mohawk Interruptus: Political Life Across the Borders of Settler States*. Duke University Press.

Simpson, Leanne. 2011. *Dancing on Our Turtle's Back: Stories of Nishnaabeg Re-Creation, Resurgence, and a New Emergence*. Arbeiter Ring.

Sivkov, Denis. n.d. "Space Exploration at Home: Amateur Cosmonautics in Contemporary Russia." Paper presented at the Anthropology Off Earth Conference, June 4, 2019, Collège de France, Paris, France.

Smiles, Deondre. 2020. "The Settler Logics of (Outer) Space." Society and Space. https://www.societyandspace.org/articles/the-settler-logics-of-outer-space

Smith, Cameron, and Evan Tyler Suliëman Davies. 2012. *Emigrating Beyond Earth: Human Adaptation and Space Colonization*. Springer.

Smithson, Michael. 2007. "Toward a Social Theory of Ignorance." *Journal of the Theory of Social Behaviour* 15(2): 151–172. https://doi.org/10.1111/j.1468-5914.1985.tb00049.x

Solon, O. 2018. "Elon Musk: We Must Colonise Mars to Preserve Our Species in a Third World War." *The Guardian*, March 11. https://www.theguardian.com/technology/2018/mar/11/elon-musk-colonise-mars-third-world-war

Spillers, Hortense J. 1987. "Mama's Baby, Papa's Maybe: An American Grammar Book." *Diacritics* 17(2): 64–81.

Spivak, Gayatri Chakravorty. 2005. *Death of a Discipline*. Columbia University Press.

Square Kilometre Array. 2016. "SKA Prospectus."

Squyres, Steven W. 2005. *Roving Mars: Spirit, Opportunity, and the Exploration of the Red Planet*. Hyperion.

Squyres, Steven W., Raymond E. Arvidson, Eric T. Baumgartner, et al. 2003. "Athena Mars Rover Science Investigation." *Journal of Geophysical Research* 108(E12): 1–21.

Star, Susan Leigh. 1999. "The Ethnography of Infrastructure." *American Behavioral Scientist* 43(3): 377–391.

Stengers, Isabelle. 2011. *Thinking with Whitehead. A Free and Wild Creation of Concepts*. Harvard University Press.

Stepan, Nancy Leys. 1986. "Race and Gender. The Role of Analogy in Science." *Isis* 77(2): 261–277. https://doi.org/10.1086/354130

Stewart, Kathleen. 2011. "Atmospheric Attunements." *Environment and Planning D: Society and Space* 29: 445–453.

Strathern, Marilyn. 1990. "Artifacts of History: Events and the Interpretation of Images." In *Culture and History in the Pacific*, edited by J. Kiikala. Finish Anthropological Society.

Strughold, Hubertus. 1959. "Advances in Astrobiology." In *Proceedings of Lunar and Planetary Exploration Colloquium*, Vol. 1, No. 6, April 25. Aerospace Laboratories.

Swift, Earl. 2021. *Across the Airless Wilds: The Lunar Rover and the Triumph of the Final Moon Landings*. Custom House.

Szolucha, Anna, Karlijn Korpershoek, Chakad Ojani, and Peter Timko. 2022. *Planetary Ethnography*. Institute of Ethnology and Cultural Anthropology, Jagiellonian University.

Taussig, Michael. 1993. *Mimesis and Alterity: A Particular History of the Senses*. Routledge.

Tejfel, Victor. 2010. "Gavriil Adrianovich Tikhov (1875–1960): A Pioneer in Astrobiology." *Highlights in Astronomy* 15: 720–721.

Terblanche, R. 2024. "*Ongedierte*: Commercial Farmers, Black-Backed Jackals and Human–Wildlife Conflict Around the Square Kilometre Array (SKA), 2016–2019." In *Contested Karoo; Interdisciplinary Perspectives on Change and Continuity in South Africa's Drylands*, edited by C. Walker and M. T. Hoffman. UCT Press.

The Exploration Museum. n.d. "The Exploration Museum." Accessed April 4, 2024. https://www.explorationmuseum.com/the-exploration-museum-3

Thompson. E. P. 1980. "Notes on Exterminism, the Last Stage of Civilization." *New Left Review* 121: 3–31. https://newleftreview.org/issues/i121/articles/edward-thompson-notes-on-exterminism-the-last-stage-of-civilization

Tikhov, Gavriil A. 1955. "Is Life Possible on Other Planets?" *Journal of the British Astronomical Association* 65(3): 193–204.

Tikhov, Gavriil A. 1959. *Sixty Years at the Telescope*. Detgiz.

Tikhov, Gavriil A. 1960. *Principal Works: Astrobotany and Astrophysics, 1912–1957*. U.S. Joint Publications Research Service.

Tikhov, Gavriil A. 1962. *L'enigme des Planètes*. En Langues Etrangeres.

Timko, Peter, Karlijn Korpershoek, Chakad Ojani, and Anna Szolucha. 2022. "Exploring the Extraplanetary: Social Studies of Outer Space." *Anthropology Today* 38(3): 9–12.

Traphagan, John W. 2016. *Science, Culture and the Search for Life on Other Worlds*. Springer.

Traphagan, John, W. 2020. "Religion, Science, and Space Exploration from a Non-Western Perspective." *Religions* 11(8). doi:10.3390/rel11080397

Traweek, Sharon. 2010. "AstroEthnography: Compiling Ethnographies/Oral Histories with Astronomers/Space Scientists." Email, January 30.

Tresch, John. 2007. "Technological World Pictures: Cosmic Things and Cosmograms." *Isis* 98: 84–99.

Tresch, John. 2012. *The Romantic Machine: Utopian Science and Technology After Napoleon*. Chicago University Press.

Tsing, Anna L. 2005. *Friction: An Ethnography of Global Connection*. Princeton University Press.

Tsing, Anna L. 2015. *The Mushroom at the End of the World*. Princeton University Press.

Tuan, Hi-Fu. 1977. *Space and Place: The Perspective of Experience*. University of Minnesota Press.

Tuck, Eve, and K. Wayne Yang. 2012. "Decolonization Is Not a Metaphor." *Decolonization: Indigeneity, Education & Society* 1(1): 1–40.

Turner, F.J. 1921. *The Significance of the Frontier in American History*. Holt.

Tutton, Richard. 2018. "Multiplanetary Imaginaries and Utopia: The Case of Mars One." *Science, Technology, & Human Values* 43(3): 518–539.

Tutton, Richard. 2021. "Sociotechnical Imaginaries and Techno-Optimism: Examining Outer Space Utopias of Silicon Valley." *Science as Culture* 30(3), 416–439.

Valentine, David. 2012. "Exit Strategy: Profit, Cosmology, and the Future of Humans in Space." *Anthropological Quarterly* 85(4): 1045–1067.

Valentine, David. 2016. "Atmosphere: Context, Detachment, and the View from Above Earth." *American Ethnologist* 43(3): 511–524.

Valentine, David. 2017a. "Gravity Fixes: Habituating to the Human on Mars and Island Three." *Hau: Journal of Ethnographic Theory* 7(3): 185–209.

Valentine, David. 2017b. "For the Machine." *History and Anthropology* 28(3): 302–307.

Valentine, David, and Amelia Hassoun. 2019. "Uncommon Futures." *Annual Review of Anthropology* 48(1): 243–260.

Van Haeringen, Maike. 2020. "Framing the 'New': Classical Reception in Amerigo Vespucci's Mundus Novus." Master's thesis, Leiden University.

van Niekerk, Piet, and Werner Hoffmann. 2019. "How Iceland Helped Humans Reach the Moon." BBC, July 2. https://www.bbc.com/travel/article/20190701-how-iceland-helped-humans-reach-the-moon

Vertesi, Janet 2015. *Seeing like a Rover: How Robots, Teams, and Images Craft Knowledge of Mars.* University of Chicago Press.

Viveiros de Castro, Eduardo. 2014. *Cannibal Metaphysics.* Univocal.

Walker, Cherryl. 2019. "Cosmopolitan Karoo: Land, Space and Place in the Shadow of the Square Kilometre Array." *Journal of Southern African Studies* 45(4): 641–662.

Walker, Cherryl, and Davide Chinigò. 2018. "Disassembling the Square Kilometre Array (SKA): Astronomy and Development in South Africa." *Third World Quarterly* 39(1): 1979–1997.

Walker, Cherryl, and M. T. Hoffman. 2024. "Contested Karoo: An introduction." In *Contested Karoo: Interdisciplinary Perspectives on Change and Continuity in South Africa's Drylands*, edited by C. Walker and M. T. Hoffman. UCT Press.

Walker, Cherryl, and J. Vorster. 2024. "Karoo 'Dorpscapes': Social Change and Continuity in Three Small Towns—Sutherland, Loeriesfontein and Vanwyksvlei." In *Contested Karoo: Interdisciplinary Perspectives on Change and Continuity in South Africa's Drylands*, edited by C. Walker and M. T. Hoffman. UCT Press.

Wasson, Christina. 2012. "Virtual Organizations." In *A Companion to Organizational Anthropology*, edited by D. Douglas Caulkins and Ann T. Jordan. Wiley.

Waterhouse, Amanda C. 2022. "A Celestial Doctrine: James Turrell, Art, and Technology in Cold War Los Angeles." *Journal of American Studies* 56(5): 729–754.

Webb, Claire I. 2021. "Gaze-Scaling: Planets as Islands in Exobiologists' Imaginaries." *Science as Culture* 30(3): 1–25.

Weheliye, Alexander G. 2014. *Habeas Viscus: Racializing Assemblages, Biopolitics, and Black Feminist Theories of the Human.* Duke University Press.

Weibel, Deana L. 2020. "The Overview Effect and the Ultraview Effect: How Extreme Experiences in/of Outer Space Influence Religious Beliefs in Astronauts." *Religions* 11(8): Article 418.

Weibel, Deana L., and Glen E. Swanson. 2006. "Malinowski in Orbit: 'Magical Thinking' in Human Spaceflight." *Quest: The History of Spaceflight Quarterly* 13(3): 53–61.

Weitekamp, Margaret A. 2004. *Right Stuff, Wrong Sex: Gender Relations in the American Experience.* Johns Hopkins University Press.

Wells-Jensen, Sheri, Joshua A. Miele, and Brandie Bohney. 2019. "An Alternate Vision for Colonization." *Futures* 110: 50–53.

Wettergreen, D., N. Cabrol, J. Teza, et al. 2005. "First Experiments in the Robotic Investigation of Life in the Atacama Desert of Chile." In *Proceedings of the 2005 IEEE International Conference on Robotics and Automation, Barcelona, Spain.* IEEE.

Whitehead, Alfred North. 1967 (1933). *Adventures of Ideas.* Free Press.

Whitehead, Alfred North. 2011 (1925). *Science and the Modern World.* Cambridge University Press.

Willerslev, Rane. 2011. "Frazer Strikes Back from the Armchair: A New Search for the Animist Soul." *Journal of the Royal Anthropological Institute* 17: 504–526.

Wilson, Kerry-Jayne. 1990. "Kea: Creature of Curiosity." *Forest and Bird* 21(3): 20–26.

Wirthlin, Morgan, Nicholas C. B. Lima, Rafael Lucas Muniz Guedes, Ana Tereza R. Vasconselos, Francisco Prosdocimi, and Claudio V. Mello. 2018. "Parrot Genomes and the Evolution of Heightened Longevity and Cognition." *Current Biology* 28(24): 4001–4008.

Wolfe, Audra J. 2002. "Germs in Space: Joshua Lederberg, Exobiology, and the Public Imagination, 1958–1964." *Isis* 93: 183–205.

Wolff, S. A., L. H. Coelho, I. Karoliussen, and A. I. K. Jost. 2014. "Effects of the Extraterrestrial Environment on Plants: Recommendations for Future Space Experiments for the MELiSSA Higher Plant Compartment." *Life* 4(2): 189–204.

Woods, D. 2014. "Scale Critique for the Anthropocene." *Minnesota Review* 2014(83): 133–142.

Wright, Jason T., and Michael P. Oman-Reagan. 2018. "Visions of Human Futures in Space and SETI." *International Journal of Astrobiology* 17(2): 177–188.

Yano, H., et al. 2017. "In-Orbit Operation and Initial Sample Analysis and Curation Results for the First Year Collection Samples of the Tanpopo Project." *Lunar and Planetary Science,* 48. https://www.hou.usra.edu/meetings/lpsc2017/pdf/3040.pdf

Young, M. Jane. 1987. "'Pity the Indians of Outer Space': Native American Views of the Space Program." *Western Folklore* 46(4): 269–279.

Zalasiewicz, Jan. 2012. *The Planet in a Pebble.* Oxford University Press.

Zhan, Mei. 2009. *Other-Worldly: Making Chinese Medicine Through Transnational Frames.* Duke University Press.

Zubrin, Robert. 2014. "To Mars, Not the Moon." *U.S. News & World Report.* https://www.usnews.com/debate-club/should-we-go-back-to-the-moon/to-mars-not-the-moon

Zubrin, Robert. 2020. *The Case for Mars.* Free Press.

Index

For the benefit of digital users, indexed terms that span two pages (e.g., 52–53) may, on occasion, appear on only one of those pages.